Democracy and Global Warming

POLITICAL THEORY AND CONTEMPORARY POLITICS

Series Editors: Richard Bellamy, University of Reading, Jeremy Jennings, University of Birmingham, and Paul Kelly, London School of Economics and Political Science

This series aims to examine the interplay of political theory and practical politics at the beginning of the twenty-first century. It explores the way that the concepts and ideologies which we have inherited from the past have been transformed or need to be rethought in the light of contemporary political debates. The series comprises concise single-authored books, each representing an original contribution to the literature on a key theme or concept.

Also published in this series:

Love and Politics: Women Politicians and the Ethics of Care
Fiona MacKay

Political Morality: A Theory of Liberal Democracy
Richard Vernon

Liberalism and Value Pluralism
George Crowder

Invoking Humanity: War, Law and Global Order
Danilo Zolo

Forthcoming titles:

Defending Liberal Neutrality
Jonathan Seglow

Seductive Virtue: The Socratic Art of Civic Education
Russell Bentley

The Politics of Civil Society
James Martin

Political Theory and the Media
Alan Finlayson

Democracy and Global Warming

Barry Holden

Continuum
The Tower Building, 11 York Road, London, SE1 7NX
370 Lexington Avenue, New York, NY 10017–6503

First published 2002

British Library Cataloguing-in-Publication Data
A catalogue record for this book is available from the British Library.

ISBN 0–8264–5071–7 (hardback)
0–8264–5070–9 (paperback)

Typeset by YHT Ltd, London
Printed and bound in Great Britain by Biddles Ltd, Guildford and King's Lynn.

Contents

To Barbara

Introduction

Global warming is a major problem of our times. Michael McCarthy, the Environment Editor of the *Independent*, commenting on the latest report of the United Nations Intergovernmental Panel on Climate Change (IPCC), says that leading atmospheric scientists insist that global warming 'is threatening the Earth with disaster' (*Independent*, 12 July 2001). It is true that there are still some who deny that global warming is taking place, or that it constitutes a serious problem. Others, while recognizing that global warming is occurring, do not accept that it results from human activities and hence effectively deny that action can be taken to combat it. The great majority of atmospheric scientists, however, now accept that man-made emissions – chiefly, but not only, of carbon dioxide – are aggravating the so-called 'greenhouse effect',[1] thereby causing the world to warm up to what amounts to a dangerous extent. According to the latest report of the IPCC, temperatures could rise by as much as 5.8°C by the end of this century. This will in all probability have very undesirable, not to say downright dangerous, consequences, such as the spread of diseases, rising sea levels and more frequent violent storms. It is increasingly being accepted that some of these phenomena are already occurring: 'Do not just think of this as the future. The effects of a warming world are already vividly visible' (McCarthy, *Independent*, 12 July 2001).

It is accepted here that global warming[2] does indeed constitute a very serious problem and that it results from human activities. It follows that something can, and ought to, be done about it.

Although it is acknowledged that global warming cannot now be prevented, it is widely (though not universally) held that it can be slowed and its effects mitigated. I shall refer to this in such terms as 'combating' or 'taking action against' global warming ('abatement' is another term that is much used in the literature on global warming).

How to bring about action against global warming is, then, a matter of urgency and of very considerable importance. Unfortunately, it is also a matter of very considerable difficulty. The reasons for this will be discussed at some length in the course of this book, but there are two main ones. First, since measures to secure effective action will probably – and many would say must necessarily – involve material sacrifices now for the sake of somewhat uncertain long-term gains,[3] there is much doubt about the readiness of people to countenance such measures. Second, effective action must, clearly, be *global*; and there are great difficulties in bringing about global action in a world of independent, sovereign states.

I shall argue that democracy provides a way of overcoming both these sorts of difficulties. But my starting point must be that there appear to be strong – and many would say compelling – reasons for thinking democracy to be, in fact, totally unsuitable for this. On the one hand, it would seem that dealing with global warming is a matter for scientific experts rather than ordinary people, and that in any case it is only authoritarian action that can disregard or overcome the unreadiness of people to countenance the necessary measures. On the other hand, since democracy is a form of state power it cannot be applicable on a global scale. I shall challenge the first line of argument by bringing forward traditional democratic arguments against government by experts and by showing that it is *only* democracy that can overcome people's initial unreadiness to support tough policies. I shall rebut the second line by taking up the idea of *global* democracy and arguing that the development of such a democracy would provide the best – perhaps the only – means of dealing with the problem of global warming.

In a little more detail, the argument will be developed as follows. First, in Chapter 1, I assess the place of democracy within green political theory. This will raise the general issue of whether or not democratic decision-making is suitable for dealing with environmental problems, and thus set the scene for the discussion of democracy and the particular – and particularly important – environmental problem of global warming. Both authoritarian and democratic strands in green political thought will be considered, and this will raise in a general form some of the key issues to be taken up in the specific context of global warming.

In Chapter 2 I challenge the argument that democratic decision-making is unsuitable for dealing with the problem of global warming since it is a matter for scientific experts. Here, an application of the traditional democratic arguments against the Platonic idea of 'guardianship' will involve some assessment of the nature of scientific expertise and its role in the global warming problem. It will also lead into a consideration of how some of the positive benefits of democracy – such as legitimation and the mobilization of consent – apply in this context. In Chapter 3 I discuss the issue, especially important in the case of global warming, of whether democracy is suitable for generating policies that promote people's long-term interests but which conflict with those that are short term. The apparent unsuitability of democracy is challenged by maintaining instead that its educative and legitimating functions are actually advantageous in securing tough policies. The interlinked issue of the extent to which democracy can prioritize policies that are in the interests of the future is then taken up, and this leads to an exploration of the idea of 'intergenerational democracy'.

In Chapter 4 I move on to a discussion of global democracy and the role it could have in generating action to combat global warming. This will involve looking in some depth at how global democracy would 'operationalize', and sharpen, the earlier argument that democracy has a positive role in generating action, and discussing how it could overcome the obstacles to action arising from the problems of global collective action and international social justice.

Finally, I will briefly raise the question of the extent to which a concern with global democracy is a realistic approach to the problem of combating global warming.

It should be made clear that there are two interconnected premises upon which the main argument of this book rests. The first, as already indicated, is that global warming is, indeed, a serious problem and action clearly ought to be taken to deal with it. The second, leading on from this, is that the judgement that such action ought to be taken is accepted as right. This may be interpreted as asserting that the judgement is in some sense objectively correct, or simply that it is here being taken as an indisputable given. (Either way, it is maintained that there is a strong rational grounding for the judgement.) The import of this premise is assessed in Chapter 2, where the possibility of popular rejection of action to combat global warming is discussed.

Notes

1. 'The greenhouse effect is a natural phenomenon, whereby certain gases in the atmosphere keep the earth's temperature significantly higher than it would otherwise be. . . . These gases allow radiation to pass through from the sun, but absorb the lower frequency, longer wavelength radiation from the earth's surface, thereby trapping heat in the atmosphere' (Paterson, 1992: 157). Such gases – notably carbon dioxide – have been increasingly emitted since the advent of the industrial revolution but 'all increased significantly during the 20th Century, both in terms of emissions and atmospheric concentrations' (Paterson, 1992: 157). For an authoritative and readable introduction to the science of the greenhouse effect and global warming, see Houghton (1997).
2. 'Three terms have been used in relation to this environmental problem: "climate change", "the greenhouse effect" and "global warming" . . . for practical purpose there is no significant difference' (Paterson, 1996: 196). In this book I shall largely stick to the term 'global warming', although it should be noted that of late 'climate change' has become more widely used.
3. The gains are the avoidance of the adverse effects of global warming. The matter of the uncertainty is commented upon further in Chapter 3, but, in brief, there remains some uncertainty about whether there really will be such effects, and just what their nature will be if they do occur; and there is uncertainty about just when and where they will occur. (In important senses the whole world will suffer, but some parts more than others; and in some areas there may even be some temporary benefits.)

CHAPTER 1

Green Political Theory and Environmental Problems

There is a general issue concerning democracy and the environment, as the title of the book edited by Lafferty and Meadowcroft (1996b) indicates. And in important respects the issues relating to democracy and the problem of global warming are a part of this. The general issue concerns the suitability of democracy for dealing with environmental problems, and it centres on the question of whether, or how far, it is feasible or desirable for responses to environmental problems to be decided democratically. Involved in this are more specific questions about whether, or the extent to which, such features of environmental problems as uncertainty, the need for expert knowledge and the length of the relevant time scales make democratic decision-making unsuitable.

These questions are the focus of the next two sections of this chapter, and help to set the scene for more detailed discussion of the questions as they arise in the particular context of democracy and the problem of global warming. Their consideration does, however, overlap with, and must be seen against the background of, some wider theoretical concerns in green political thought. There is, indeed, a general issue concerning democracy in green political theory,[1] where the question of the suitability of democracy for dealing with environmental problems becomes a part of more general green theoretical concerns.

It is perhaps true to say that broadly speaking there is a commitment to democracy in contemporary green political theory (though not all would agree, and I shall take this up below). In part this reflects the current status of democracy: today 'one can posit democracy as a value to be considered as an essential part of all acceptable political theories. In this respect green politics is no different in its claim to be part of the "democratic project"' (Barry, 1999: 193).

I shall look at this idea of contemporary green political thought being committed to democracy shortly. First, though, we should realize that there have been anti-democratic strands in green theory.

Anti-democratic strands in green political theory

It must be recognized that at least in the past there has been a clear anti-democratic element in green political theory. 'Amongst those writing from an ecological point of view', it can be said, in fact, that 'there have been two standard and contradictory responses to the relationship between green politics and democracy. From green parties and radical green movements has come a stress on the need for participatory democracy', while by 'contrast from "survivalists" has come a contradictory argument that sees democracy as an obstacle to dealing with ecological crisis' (Doherty and de Geus, 1996a: 1). Such 'survivalist' theory was prominent in the 1970s. For Robyn Eckersley 'ecopolitics' has, roughly speaking, evolved through three stages since the 1960s:

> the initial stage . . . was 'participatory' and grew out of the protest and 'new left' politics of the 1960s. In the following decade, however, the primary theme became 'survivalist', with democratic commitments sacrificed to the overriding concern of preventing ecological catastrophe. In the 1980s survivalism gave way to a new 'emancipatory' environmentalism that is still growing and developing. (Taylor, 1996: 86, referring to Eckersley, 1992)

With global warming frequently being regarded as leading to an ecological catastrophe, it might seem that survivalism would be the appropriate stance. However, for various reasons that are taken up in the course of the argument of this book, this view is to be rejected in favour of a democratic stance. The survivalist stage in green political thought was, indeed, characterized by strongly anti-democratic thought, an important theme in which was that 'ecological crisis could be tackled only by a strong government that would be prepared to curb the freedom of individual citizens' (Doherty and de Geus, 1996a: 1). This kind of thinking was particularly associated with the 'terrible trio' (Barry, 1999: 195, 243) of Ophuls (1977), Hardin (1968, 1977) and Heilbroner (1974). The message was that 'only authoritarian regimes have the power required to compel individuals to refrain from consuming too many resources; without such compulsion, the survival of human society, perhaps life itself, is very much in question' (Taylor, 1996: 87). Here, then, is 'ecoauthoritarianism'; and it provided, 'during the 1970s, some of the most frankly authoritarian thought of the contemporary era' (Taylor, 1996: 87).

An important aspect of this 'ecoauthoritarianism' is a variant of the traditional anti-democratic argument, from Plato onwards, that the mass of the people lack the necessary knowledge. I shall take up this argument in its particular application to the problem of global warming in the next chapter, but let me note here its general form. Barry says of Ophuls that the 'justification of his anti-democratic stance is basically the traditional argument of "the ship of state" requiring the best pilots, and the dangers of "rule by the ignorant" when faced with such a complex and complicated issue as social-environmental dilemmas' (Barry, 1999: 195). This 'technocratic' dimension of ecoauthoritarianism (Barry, 1999: 195) is sometimes supplemented by a 'metaphysical' dimension. Barry (1999: 200–1) refers to

> Those analyses of ecological issues where solutions must partake of some definitive (often metaphysical) 'truth' in respect to social-enviromental relations. From such perspectives a 'proper' social-environmental metabolism is derived not from objective scientific principles, as in the technocratic-authoritarian argument . . . but from the revealed truth of some metaphysical or spiritual system of belief.

This dimension is not, perhaps, overtly acknowledged in debates over global warming but it does sometimes have an influence. Some of the more committed proponents of drastic action to curb global warming are (or are influenced by) green theorists who have what seems to be a metaphysical conviction that anthropogenic global warming is occurring, that it is leading (or will lead) to an environmental catastrophe and that drastic remedial action is essential. At any rate, their opponents see it as a metaphysical conviction, since they accuse such proponents of ignoring, or not basing their views on, careful weighing of scientific evidence. The issues here are complex, not least because of some (though now diminishing) dispute regarding the scientific evidence and, indeed, regarding the status of science itself. I shall take up such issues in the next chapter.

Besides the argument from the requirement of (scientific or metaphysical) knowledge of the truth, ecoauthoritarians also tend to stress the incapacity of the mass of the people to make the necessary sacrifices to obtain ecological goals. The argument or assumption here is that the management of ecological crisis entails the kind of drastic reductions in material prosperity to which people would not ordinarily consent. 'Given "human nature" (which Heilbroner saw as fundamentally selfish), our only hope for survival lay in our obedient rallying behind a centralized, authoritarian nation – the only institutional form that Heilbroner saw as capable of extracting the necessary sacrifices' (Eckersley, 1992: 14; referring to Heilbroner, 1974). I shall take up the issue of the capacity of the mass of the people to make the sacrifices required for successful action to curb global warming in Chapter 2. But we should notice here that the question is not just one of selfishness, since equally important is the question of people's capacity to sacrifice their short-term interests for (their own, as well as other people's) long-term interests.

It might be argued that the problem of scale reinforces such survivalist ecoauthoritarian arguments, since it is commonly thought that it is the scale as well as the urgency, severity and complexity of ecological problems that puts their solution beyond the capacity of democracy. The scale of many ecological problems does, indeed, spread them beyond the control of existing democratic governments. But the ecoauthoritarian argument that 'only a concentration of power will be effective enough to achieve [their] solution' (Doherty and de Geus, 1996a: 11) is irrelevant here since the scale of such problems – above all, of course, the problem of global warming – puts them beyond the control of authoritarian states too. Indeed, as Doherty and de Geus point out (1996a: 11), it should be said that here are 'problems of the effectiveness of nation-states rather than necessarily problems of democracy'. The authoritarian argument could only be bolstered by positing the viability of that which is widely regarded as unfeasible and/or undesirable, namely supra-state concentrations of power. In fact it is a crucial argument of this book that it is democracy that is best able to cope with the supranational scale of environmental problems (in particular the problem of global warming), albeit democracy of a new kind. This is because democracy – this new kind of democracy – can be supranational. To put it another way, although a fundamental failing of democracy as traditionally understood and practised is its inability to cope with supranational environmental problems, this *is* a failing of democracy of this traditional kind rather than of democracy as such.

But let us return to the general nature of the anti-democratic strands in green political theory. Outright ecoauthoritarianism did not last long. Taylor (1996: 87), for example, asserts that 'this mode of environmental thought was relatively short-lived', and Dryzek (1996: 108) maintains that '[i]f two or more decades of political ecology yield any single conclusion, it is surely that authoritarian and centralized means for the resolution of ecological problems have been discredited rather decisively'. And contemporary green theory may be seen as predominantly democratic.

Some, however, do highlight remaining anti-democratic elements. Even where it is not unambiguously authoritarian, green theory may, arguably, be pulled in that direction (further ambivalence regarding democracy will be discussed below). A salient argument here is developed by Saward, to the effect that, even if democracy is seen in some sense as desirable, 'it is simply incompatible with other green goals' (Mills, 1996: 98; referring to Saward, 1993). Saward's argument is that, even if democracy can in some ways be seen as a green value, or as being compatible with some green values, sometimes it is necessarily going to be in conflict with green goals; and it is these goals to which greens will give priority where there is such a conflict. Saving the environment is the dominant imperative and if, or where, this cannot be done by democratic means, it must still be done: if the people do not want to save the environment, it must nonetheless still be saved. Taylor makes a similar point by referring to Westra (1994) with 'her emphasis on right-thinking "governments, professionals, and institutions" who understand the truth, are willing to attempt to educate the public, but who are willing to do what is right regardless of public opinion' (Taylor, 1996: 97).

It is true that Saward (1996) subsequently qualified this argument. He first affirms the incompatibility thesis: 'If democratic rule is responsive rule, then (to put it bluntly) the majority should get what it wants. If it does not want (vote for) green outcomes, so be it' (83). But, taking up the argument that in a genuine democracy majority rule is limited by democratic rights (of individuals and minorities), he then considers whether 'at least some substantive environmental concerns' might not 'be built into the very structure of democratic theory via a "green democratic right"' (84). However, he concludes that environmental imperatives cannot necessarily be contained within democratic rights and reiterates that to 'the extent that green outcomes take precedence over strictly democratic outcomes, it ought to be recognized and acknowledged that democracy is being diluted' (93).

A similar conclusion is reached by Goodin, who, as Dobson points out, 'poses the problem succinctly: "To advocate democracy is to advocate procedures, to advocate environmentalism is to advocate substantive outcomes: what guarantees can we have that the former procedures will yield the latter sorts of outcome?"' (Dobson, 1996: 133–4). Taylor makes much the same point about Naess, despite the democratic aspects of his thought (Taylor, 1996: 96). He concludes: 'Push comes to shove: Naess's commitment to a particular set of policies, and the ideology from which they derive, clearly override his commitment to democracy' (Taylor, 1996: 97; referring to Naess, 1989).

The force of the argument that the achievement of green goals may conflict with democracy is brought home by reflection on the nature of some of the goals (or, more precisely, the nature of the goals to be found in some varieties of green political thought). I have already remarked on the ecoauthoritarian idea that the necessary economic sacrifices require strong government. But even without prescribing authoritarian solutions one can see there is at least a problem with regard to democracy here. According to radical green analyses, in order to deal properly with ecological problems – and this would be just as true of those which are global – and obtain a sustainable society, 'radical changes in our social habits and practices are required' (Dobson, 2000: 16). Such changes would, indeed, amount to de-industrialization and would involve (what according to 'conventional viewpoints' would be seen as) drastic reductions in standards of living, at least in advanced industrial states. Now, it is easy to argue that such changes are not wanted by the vast majority of people – and that policies or proposals embodying or advocating them would be deeply unpopular – and hence that they could not be brought about by democratic means. Robert Cox sums up this view as follows:

> It would seem that a radical change in patterns of consumption will become essential to maintenance of the biosphere. . . . [A] change of lifestyle may be necessary to biospheric survival and at the same time [it is acknowledged] that political survival in modern liberal democracies makes it highly risky for politicians to advocate. . . . Is biospheric survival incompatible with liberal democracy? Will democratic politics continue to press for maximizing consumerism? (Cox, 1997: 67–8)[2]

It is apparent, then, that an incompatibility between democracy and green goals can readily be discerned. It is possible to see this incompatibility in terms either of the achievement of green goals requiring undemocratic means, or of democracy blocking the achievement of such goals. And a central strand in the argument of this book will, in effect, consist in a testing out of this incompatibility thesis in respect of the particular case of global warming.

The commitment to democracy in later green political thought

It does seem, then, that there is, at the very least, a crucial anti-democratic element in green political thought. But how true is this in fact? We have already seen that contemporary green political theory has been characterized as essentially having a commitment to democracy, so there must be a different way of viewing the relationship between democracy and the achievement of green political goals. I will now consider this alternative stance.

I remarked earlier on ecoauthoritarianism being seen as but a stage in green political thought, so that 'the green movement in its modern form can confidently be said to have abandoned authoritarian solutions to the environmental crisis' (Dobson, 1995: 27). There was, in fact, 'widespread criticism of the authoritarian response' (Eckersley, 1992: 17)[3] and a growing perception that democracy was in fact *better* than authoritarian systems at achieving green goals. One important reason for this perception was that the collapse of communism revealed that authoritarian regimes had a *worse* environmental record than liberal democracies: 'there is now ample evidence that, however ineffective democratic regimes have been regarding environmental protection, they have been far more effective than were the authoritarian regimes of the former Soviet Union and Eastern Europe' (Paehlke, 1996: 19).

Besides this highlighting of the negative record of authoritarian regimes there has also been increasing emphasis on the positive reasons for the ecological superiority of democracy. Paehlke (1996: 19) provides a good summary statement:

> Closed, technically based, decision making regarding an objectively determined public interest was . . . rejected . . . in favour of wide democratic participation. Wide participation is seen as necessary to determine essentially subjective and value-laden environmental policy objectives. Only a participatory approach to policy making can incorporate the needs of all segments of society, future generations and other species.[4] Environmental values and their policy making implications are best understood if all segments of society put forward their own views for themselves. The views of scientific and technical experts . . . are, in this view, but one (or two) voice(s) among many.

This statement includes a counter to the fundamental ecoauthoritarian argument concerning the inability of the mass of the people to discern ecological truths. This argument about the incapability of the mass of the people is discussed in some depth, in relation to the particular issue of global warming, in the next chapter. Indeed, most of the other issues raised in the following paragraphs are also pursued further, in this context, in the next chapter. It is, however, illuminating to raise these matters initially in the wider context of green political theory. And the counter-argument here to the incapability-of-the-masses thesis is that pursuing green goals is a matter of determining 'essentially subjective and value-laden policy objectives'. However, green democratic theory does not necessarily accept this relativist stance and other counter-arguments can be deployed. Dobson, for instance, points out that greens can argue that democratic decision-making is 'the method by which ecological truths will be revealed' (Dobson, 1996: 140). He puts the argument this way:

> To the degree that that there is a determinate answer to the 'right' values and the 'right' kind of society in which to live

> (and the greens, in the round, believe there is), then greens should be committed to democracy as the only form of decision making that – for Millian reasons – will necessarily produce the answer. (Dobson, 1996: 139)

There is a parallel pro-democracy green argument premised on uncertainty rather than determination of the truth. 'Democracy can . . . be defended as the most appropriate collective decision-making procedure under conditions of uncertainty' (Barry, 1999: 203) – and comprehending and responding to ecological problems is, indeed, typified by such conditions. This can be seen as a matter of obtaining necessary support: 'Only a democratic, "open society" can hope to make good (as opposed to "true") decisions regarding material interaction between society and its environment which can command widespread support' (Barry, 1999: 204). But – and this is closer to the determination-of-the-truth argument – it can also be seen in Popperian terms: 'For Beck, following Popper to some extent, democratic decision-making is a form of institutionalized self-criticism which he sees as "the *only way* that the mistakes that would sooner or later destroy our world can be detected in advance" ' (Barry 1999: 204; quoting Beck, 1992: 234, emphasis in the original).

There are other somewhat similar or overlapping arguments which focus on the nature of scientific enquiry. These will be taken up in the next chapter (where the previous arguments will also be considered in more depth), but we should note here the important argument that 'since scientific knowledge and its technological application can have effects on individuals . . . those affected ought to have some say in how science is used' (Barry, 1999: 203). Indeed, a 'green commitment to democracy can be understood as expressing a desire that technological and scientific development be subject to the ultimate authority of the demos' (Barry, 1999: 205).

The issue, then, is not just one of the demos having a say in scientific matters but, rather, one which embodies a central theme in green thought, namely, the desire to reverse – or at least to gain some control of and to modify – the techno-industrial development of modern society. Barry (1999: 205) refers to the 'green suspicion of technologically led economic growth which is largely beyond democratic control but which has great and far-reaching effects on the citizens of the demos'. This ties in with the democratic implications of the green notion of sustainability.

In green thought conventional notions of techno-industrial economic development, based on continuing exploitation of the environment, are typically replaced by the ideas of sustainability and sustainable development. And, arguably, these are ideas which pose issues that require resolution by democratic decision-making.

Sustainability and sustainable development are ideas which are of growing importance. Clearly, they are of central importance in green political thought. And they are also – at least in terms of the lip service often paid to them – growing in influence beyond this. Indeed, '[s]ince the publication of *Our Common Future* (WCED, 1987), sustainable development has become internationally the principal aim of environmental policy' (Achterberg, 1996: 157). But just what does sustainable development amount to[5] and how is it to be realized? Answering such questions involves judgement and debate. To use the Bruntland Commission's definition, '[s]ustainable development is development that meets the needs of the present generation without compromising the ability of future generations to meet their own needs' (WCED, 1987: 9). But this raises crucial issues about, for example, what are to be included as 'needs' and how far – how many generations – in the future we should look. And this suggests an important link between sustainable development and democracy.[6] Answering the questions involved here requires judgement rather than appeals to experts for the 'true' answers. And, following traditional justifications of democracy, it is the mass of the people who should in the end make such judgements, not least because they will be importantly affected by them. As Barry (1999: 206) puts it:

> The indeterminacies thrown up by sustainability require political adjudication, and, given that policies flowing from any conception of sustainability are likely to have a widespread social impact, leaving few citizens' lives untouched, it is uncontroversial to hold that they should have some say in its articulation and formulation. That is to say the indeterminacy of the principle calls for citizen deliberation, while its translation into policies and laws calls for their consent and, equally important, their participation in achieving it.

Different forms of democracy

I have now indicated some of the main lines of argument in support of democracy to be found in green political theory. It is not, however, always, or necessarily, democracy *per se* that is supported, so much as particular forms of democracy.

One indication of this is the assessment of *liberal* democracy. We shall see below that my earlier suggestion that there is a commitment to democracy in contemporary green political theory may be something of an oversimplification. There is in fact some ambivalence. But this is partly, or sometimes, a matter of ambivalence concerning liberal democracy rather than democracy as such. This point can be obscured because today liberal democracy is often seen as the only – or the only viable or valid – form of democracy,[7] and therefore doubts about liberal democracy can be construed as doubts about democracy itself.

Before going any further I need to clear up a confusion arising from an uncertainty, or looseness, in the meaning of 'liberal democracy'. Green concerns about liberal democracy (which I shall indicate in a moment) usually involve distinguishing it from – and contrasting it unfavourably with – other forms of democracy. And, indeed, the point of the previous paragraph depends on making this kind of distinction. But this sits oddly with the dominant view that liberal democracy is the only kind of democracy there can be. Now, it may be that green critics of liberal democracy are challenging this dominant view. Sometimes this may be the case;[8] but normally it is not so. Rather, it is usually a case of 'liberal democracy' being understood in a wider or narrower sense; but this is seldom made clear because the use of two meanings is seldom made explicit. On the one hand, in the wider, generic, sense 'liberal democracy' roughly speaking refers to any system in which governmental decisions are the outcome of, or are subject to, free, competitive elections based on universal adult suffrage. And this is the meaning employed when liberal democracy is viewed as the only kind of democracy there is. On the other hand, there is a narrower, specific, sense where liberal democracy is but one of the forms this democracy – liberal democracy in the wider sense – can take.[9] In this sense liberal democracy can be contrasted with other forms of democracy (other forms of liberal democracy in the wider sense).

Green doubts, then, are typically about liberal democracy in the narrower sense and can be expressed and developed in terms of an unfavourable contrast with other forms of democracy, such as deliberative, republican or participatory. Broadly speaking, liberal democracy in this sense is seen from the perspective of its 'radical critique' (Holden, 1993: ch. 3) and is characterized as a system in which 'decisions by the people' are arrived at by means of an aggregation of purely individual decisions.[10]

There are two main types of green concern about liberal democracy in this sense (which overlap). First, there is the perceived association with capitalism and second there are problems with key implications of the liberal individualism by which it is characterized.

It is, of course, widely held that liberal democracy is linked with capitalism, and its left-wing critics commonly maintain that this linkage systematically frustrates its aims[11] (there is an extensive literature here, but see, for example, Holden 1993: 141–50, 173–7). Green critiques of liberal democracy tend to subscribe to this view and thus to see this necessary connection with capitalism as rendering it inimical to the realization of green values (here there is an overlap with general green critiques of capitalism and industrial society). Dryzek (1996: 112), for example, writes that:

> There are reasons to suppose that the limits of the liberal democratic state are very quickly reached in ways not generally recognized by . . . liberal democrats. Foremost among these reasons is the simple fact that pursuit of both ecological and democratic values becomes problematical as soon as it encounters the imperatives facing any state (liberal democratic or otherwise) in the context of a capitalist market economy. The first task of any such state is to maintain levels of economic activity.[12] . . . Policies that pursue ecological values are obvious candidates for punishment.

But it is not just the association with capitalism that is perceived as the problem; rather it is the kind of individual behaviour that is intrinsic to liberal democracy. As Dryzek (1996: 112–13) goes on to say,

> even setting aside problems that arise with the intimate relationship of the liberal democratic state with the capitalist market economy, there are features intrinsic to liberal democracy which prevent the generation of truly effective solutions to ecological problems . . . [P]olitics in liberal democracy is mostly about the pursuit of self interest.

Another way of putting the point is to say that liberal democracy erases citizenship, so that in a liberal democracy individuals behave as consumers rather than as citizens. The underlying point is made in Pateman's (1985) critique of the liberal democratic conception of the act of voting as an isolated, and private act inimical to the proper communal form of a citizen's activity with its focus on the common good. This is developed in the context of green theory when, for example, Barry (1999: 65–6) refers to

> the idea that individuals as *consumers* have interests which are different to those they have (or potentially have) as *citizens*, and that on the whole, 'ecological interests' are not well served by the former. . . . In other words, qua consumers in the market, individuals have a narrower set of ecological interests than would be the case qua citizens[13] [emphasis in the original].

Barry also refers to Sagoff's (1988) argument to the same effect:

> individuals as consumers are guided mainly by considerations of their own interests, whereas as citizens they have to, or ought to, place the latter in the context of a common good, which accommodates the interests of others as well as their own. (Barry, 1999: 208)

The implications of these green concerns about liberal democracy for the issues relating to democracy and global warming are ambivalent. On the one hand, global democracy is not properly comparable with liberal democracy (in the sense that is relevant here). There are some complex issues here, which are taken up in Chapter 4, but it is largely a matter of global democracy being a non-state form of democracy and the extent to which it is postulated as a counter to the narrower economic imperatives involved in globalization.[14] On the other hand, green concerns are one of the essential bases for the body of opinion pressing for action against global warming and this inevitably affects thought about the nature of democracy and its role in generating such action (clearly, though, since alarm about global warming is no longer restricted to green thinking, this body of opinion, and any ideas about the nature and role of democracy, also have other and wider bases).

Green responses to these concerns about liberal democracy (in the narrower sense) typically involve showing how they are met by, or do not apply to, other models of democracy – that is to say other models of liberal democracy in the wider, generic, sense. As Eckersley (1996: 217) puts it, 'by the time green political theory began self-consciously to develop in the late 1970s' it looked for '"stronger" forms of democracy than liberal democracy'.

Concerns about liberal democracy's erasure of citizenship, for example, can be countered by focusing instead on the republican model of democracy, where citizenship, and the elevation of civic virtue over individual interests, is central.[15] According to Barry (1996: 231) '[c]itizenship, as viewed by green democratic theory, emphasizes the duty of citizens . . . in the collective enterprise of achieving sustainability. There is thus a notion of "civic virtue" at the heart of this green conception of citizenship.' Again, there are no straightforward implications for ideas concerning global democracy, where, without a state, the usual notion of citizenship is inapplicable. As against this, though, two considerations are important. First, as discussed in Chapter 4, the notion of a global community is important in ideas about global democracy. And such a notion reinforces and is reinforced by a growing interest in notions of 'global citizenship'.[16] Second, as I shall also discuss in Chapter 4, a significant benefit of global democracy in relation to the problem of combating global warming would be its contribution to overcoming collective action problems by its focus on a global common good. Here there is a parallel to the republican conception of citizenship's elevation of communal concerns over individual interests.

The discounting of the expression of individual preferences and interests in favour of an emphasis on the community and the common good is more explicitly developed in the tradition of participatory democratic theory (for an overview see Holden, 1993: 122–40). And, indeed the 'decision-making rules and frameworks that have been typically defended in green movements and party circles [since the late 1970s] have usually followed a string of variations on a participatory democracy theme' (Eckersley, 1996: 217). A recent variation is the deliberative model of democracy[17] and '[t]here is a good deal of support in the literature for [the idea] that the view of democracy which best fits with green politics is a communicative or deliberative model' (Barry, 1999: 214–15). Deliberative democracy can be contrasted with liberal democracy (narrow sense) by saying that 'in the former, citizens' preferences are forged through a process of structured debate focused on the need to realize the common good, while in the latter, unrefined and perhaps uninformed preferences are merely counted up to produce public policy' (Saward, 1998: 64). Eckersley (1996: 217) puts the point this way:

> More deliberative, pedagogical forms of free and impartial public debate have been defended as being more conducive to securing long-range environmental protection[18]. . . .Deliberative or 'discursive' democracy is defended as superior because it seeks to educate through dialogue and transform political opinion through reasoned debate, rather than simply aggregate the sum of unchallenged individual desires.[19]

A notable feature of deliberative democracy here is the subsumption and transformation of purely individual preferences in the generation of shared views about the communal good. It is held that this flows from the imperatives of a system in which what is required, to reach an agreed policy for achieving the public good, is rational dialogue with other individuals. Barry (1999: 216), for example, says that the 'choice of a discursive form of democracy within green political theory is closely associated with the public goods character of social-environmental issues'. He goes on to say, quoting Jacobs (1997):

> As Jacobs notes, environmental goods are not private: 'Forming attitudes to them is therefore a different kind of process from forming attitudes (preferences) towards private goods. It involves reasoning about other people's interests and values (as well as one's own)' (1997: 219). This is at the heart of the green democratic argument [for] . . . a deliberative or discursive political process. (Barry, 1999: 216)[20]

It is important to realize that there are in fact two very different problems that are seen as being addressed here by deliberative democracy. One of these is, indeed, that of dealing with, or overcoming, individuals' perverse selfishness or short-sightedness. But the other is crucially different, though the difference is often obscured by a superficial understanding of what is at stake. The difficulty is that it is often the case that individuals are, in effect, *prevented* from doing anything other than that which would otherwise be seen as selfish or short-sighted. Indeed, the real problem here is that individuals' actions or attitudes that may appear as short-sighted and irrational are actually just the opposite. What we have, in fact, is a classic instance of the collective action problem of policies which are collectively rational (those which promote the public good) requiring actions which are individually irrational. This is often conceptualized in terms of 'the prisoner's dilemma', and in the field of the environment was given a classic expression as 'the tragedy of the commons' (Hardin, 1968). This problem will be taken up in Chapter 4, but we should note here a particular implication for deliberative democracy.

The point is this. The institution of government is the classic response to the collective action problem, whereby it becomes rational for individuals to do that which promotes the public good because governmental coercion ensures that they will *all* do it. But it is arguable that this is unnecessary,[21] or, alternatively that it will only work in an acceptable and democratic way, where individuals properly understand the situation and through communication come to coordinate their action. Liberal democracy lacks both of these necessary conditions. (I am here talking of the critics' characterization of liberal democracy, in the narrow sense, rather than necessarily what really happens in actually existing systems called liberal democracies.) The first requires that individuals focus on environmental public goods rather than narrow economic interests (Weale, 1999: 38–46), whereas it is the latter focus that is incorporated in liberal individualism.[22] The second requires a process of communication and deliberation rather than the simple aggregation of primarily self-interested individual votes characteristic of liberal democracy. In deliberative democracy, on the other hand, both conditions are met. Barry (1996: 124–5) explains this point well:

> Enhanced democratic institutions which stress citizen activity and deliberation on collective issues are more likely to avoid the prisoner's dilemma (Elster, 1983) with regard to environmental problems. Communicative rationality makes it less likely that the collective results will be ecologically irrational. The ecoauthoritarians' formulation of the paradigmatic ecological problem in terms of the 'tragedy of the commons' can be criticized therefore for not making any allowance for purposeful communication between individual users of the commons. It simply assumes a prisoner's dilemma scenario with mutually disinterested and non-communicating 'rational individuals'. However, by introducing a communicative dimension, an intersubjective realm is created which permits the co-ordination of individual activity in such a way that the aggregate effect of individual behaviour is not, as in the tragedy scenario, both collectively and individually undesirable.

One further feature of deliberative democracy, noted by Barry, also stems from the way in which individual preferences are superseded or transformed in the process of deliberation. 'Raw' individual preferences are not the product of critical, rational thought so much as simply reflections of existing non-ecological culture – rather as for Marxists individuals' preferences in 'bourgeois democracies' are the product of 'false consciousness'.[23] Only deliberative democracy, therefore, will bring people to think critically about the status quo and to espouse ecological values. This argument is of critical importance when it comes to attitudes towards (quite possibly economically damaging) action against global warming.

As suggested above, deliberative democracy is a form, or understanding, of participatory democracy. Participatory democracy is not the same as, but is inspired by, direct democracy (Holden, 1993: 136–8); and it might seem that green support for deliberative democracy would be linked with a predilection for direct democracy.[24] But such a predilection, while seen as characteristic of green participatory democratic theory, is not *necessarily* a part of green support of deliberative democracy. Barry (1996: 120), for example, refers to 'the common perception that green democratic theory must be some variation of direct democracy', but maintains that 'representative forms of democracy are perhaps more central to green concerns than is usually thought'. One reason for this is that, arguably, 'direct democracy works best where there is already a large degree of agreement between participants, and that representative forms are more suited to pluralist and heterogeneous collectivities' (121). Hence 'it seems more realistic to assert that green democracy implies that representative institutions will be supplemented by participatory democracy, rather than transcended by a direct democracy' (121). Additionally, of course, representative forms could be said to be more appropriate for global democracy, which is on a *very* large scale and involves a very substantial degree of heterogeneity. Barry concludes that green advocacy of '[d]eliberative democracy should not . . . be interpreted as a demand for direct democracy, as opposed to more participatory practice, where representative institutions can be supplemented with . . . greater citizen involvement' (122).

There is an additional reason – and one that is recognized by Barry (1999: 221) – for focusing on representative forms. This is that environmental problems – and global warming is a prime example – affect future generations as much as, or more than, the present generation. Arguably then, they too should in some sense have a say in present decisions on environmental policies. The only way this can be done is via representation, in some sense, of future generations. This could be taken to imply an 'intergenerational democracy' – which, clearly, could only be a representative form of democracy. These issues are taken up in Chapter 3.

Arguments for representative forms of democracy are, then, to be found in some strands of green political theory, and these counter the commitment to direct democracy found in other strands. On the other hand, though, such a commitment is often modified in a contrary direction, and 'radicalized' into 'eco-anarchism' (Eckersley, 1992: ch. 7). 'The characterization of green politics as . . . anarchistic . . . has wide currency' (Barry, 1999: 77). And there is an 'almost complete monopolization of the green political imagination by an anarchist vision of the society greens would like to create' (Barry, 1999: 77). Thus the 'classic ecocentric proposal is the self-reliant community modelled on anarchist lines' (O'Riordan, 1981: 307). But a stateless society – which is the basic anarchist notion – has somtimes been held to be the most pure form of direct democracy (Wolff, 1970),[25] so that not only does 'eco-anarchism [express] the green concern for democracy' (Barry, 1999: 3) but it does so by embracing what is seen as the most thoroughgoing form of direct democracy.

Now, such eco-anarchist ideas besides being anti-statist are focused on small-scale communities. This is a manifestation both of the 'small is beautiful' (Schumacher, 1976) precept and of a concern with the importance of community. However, today small homogeneous communities hardly seem the appropriate sites for ecological action: certainly many of the most serious environmental problems – and above all the problem of global warming – require action on a large scale and involving diverse segments of society and, indeed, diverse societies. And the anti-statist aspect of eco-anarchism is not necessarily limited to a focus on the small homogeneous society. Barry (1999: 236) argues for eco-anarchism to be superseded by the 'incorporation of the concept of civil society into green political theory'. And, tapping into the contemporary concern with the role of civil society in democracy (which I take up in Chapter 4) he argues for a shifting of power from the state to civil society (Barry, 1999: 238). Here is an anti-statist stance that does not focus on small-scale communities and which incorporates a concern with heterogeneity. These ideas are developed further in

Chapter 4, where I discuss global civil society and global democracy.

Before I leave this outline of the nature of green commitments to democracy we should note that there is to an extent a dichotomy in green democratic theory between instrumental and intrinsic commitments to democracy. As Eckersley (1996: 223) points out there is a difference between judging democracy 'to be the best available means of securing green goals' and seeing a 'connection between ecology and democracy [that is] no longer contingent. That is, authoritarianism would be ruled out at the level of green principle' (see also Dobson, 1996 for an important discussion of this issue). In the case of democracy and global warming our concern is with the instrumental arguments – to the effect that democracy is the best decision-making system for combating global warming. Nonetheless, it should be remembered that the traditional reasons for valuing the democratic process – the traditional reasons for maintaining that democracy is the best system of government – are also involved. Indeed, they overlap with, and interpenetrate, the instrumental arguments. For example, the case for popular rather than purely 'expert' assessment of the global warming issue and the proper response to it, centrally involves the traditional arguments of democratic theory against what Dahl (1989) calls 'guardianship'; and I shall consider these in the next chapter. Also – another example – the other side of the coin of the vital 'legitimacy argument' (taken up in Chapter 3), concerning the need for decisions imposing material sacrifices to be taken democratically if they are to stick, is that to be morally justified such decisions must only be taken with the consent of those whom they affect.

Green ambivalence towards democracy?

Whether for instrumental or intrinsic reasons it seems, then, that there is today a firm commitment to democracy – even if not to liberal democracy, as narrowly understood – in green political thought. Indeed, it is common for this connection now to be taken for granted, so that '[d]emocracy and enhanced environmental protection have been taken to be self-evidently mutually reinforcing' (Lafferty and Meadowcroft, 1996a: 2).

However, the situation is not really quite as clear-cut as this. Some doubts about democracy do remain among some green theorists; or, to put it another way, some problems are still discerned regarding the suitability of democracy for dealing with environmental problems. Lafferty and Meadowcroft (1996a: 2), for instance, warn that '[i]n fact there are good reasons for believing that the relationship between democracy . . . and good environmental practice . . . is far from being straightforward'; while Taylor (1996: 102) maintains that 'despite encouraging correlations between democracy and environmentalism, it is important to remember that there is no simple or uniform relationship between the two' and 'environmentalism as a whole is neither democratic nor undemocratic, although particular movements and theorists can certainly be seen to be one or the other'.

This may be to overstate the case. Even where there is a sharpened appreciation of the problems that remain, the predominant view now does seem to be that there is a linkage between green political theory and democracy. Or, to put the point another way, that there is a linkage between environmental protection and democracy, such that democracy enhances protection of the environment or, at the very least, that environmental protection and democracy are not incompatible (and I shall be examining the validity of this view in the case of global warming). The position is illustrated by the views of Robert Paehlke, who subsequently qualified the view he took in his initial rebuttal of the ecoauthoritarians:

> I still hold the views which I put forward in my 1988 article [Paehlke, 1988], but I now also see another side to the 'green democracy' coin. . . . I think now that in my haste to reject the bleak and perhaps even dangerous views of Heilbroner, Ophuls and others I did not directly enough face the many challenges to democracy and environmental protection that are frequently bound up one with the other. (Paehlke, 1996: 20)

Even so, he insists that '[o]verall, I remain steadfast in my view that environmental protection requires democracy' (Paehlke, 1996: 20).

Of central importance here is that a contributory reason for Paehlke's ambivalence is the new threat of global warming: 'many of the apocalyptic visions of the 1970s have been largely rejected (though we have "gained" new fears, such as climate warming)' (Paehlke, 1996: 21). And, of course, besides the critical 'new' issue concerning largeness of scale and the applicability of democracy to the international realm, the problem of combating global warming raises many of the issues of economic growth or material well-being that have troubled green approaches to democracy. And for Paehlke such issues continue to cause some trouble. The global warming problem, then, renews or extends the debate about democracy and environmentalism, or environmental protection. And besides being centrally concerned with the practical question of whether democracy can help promote action to combat global warming, the argument of this book can be seen as a contribution to this debate.

Notes

1. A distinction quite commonly drawn between reformist 'environmentalism' and radical ecologism (Dobson, 2000) might seem relevant here. Thus in an 'environmentalist' approach one might say that the issue of democracy and the environment consists of particular questions which are to be considered in the context of an empirical analysis of the world as it is, whereas for 'ecologism' it is a basic theoretical issue requiring general theoretical analysis, the conclusions of which are to be used in a critical evaluation of the world as it is. In these terms the discussion of global warming and democracy in this book falls predominantly within the environmentalist approach. However, the usefulness or validity of the environmentalism/ecologism dichotomy can be challenged (Barry, 1994, 1999), and at the very least there can be an important overlap between discussions of democracy and the environment within 'environmentalism' and 'ecologism'. In any case there is no standardization of such terminology within the literature: often, for example, the term 'environmentalism' is used in a way that covers either or both of the meanings of the separate terms as they are presented in environmentalism/ecologism dichotomy.
2. The next sentence reads: 'Or may people persuade themselves to refashion their aspirations so as to make them more compatible with biospheric survival?' (It may indeed be that the global warming problem itself could prove influential in such persuasion.) Moreover, many greens would maintain that because such refashioned aspirations would be in accord with a proper understanding of the world, such 'persuasion' should be easy. (This ties in with the postulate of this book that because a proper understanding of the world shows that action against global warming is indeed necessary then people will be persuaded that this is so: see the Introduction above and the remarks in Chapter 1 below.) It should, of course, be noted here that an essential feature of important varieties of green political thought is that reductions in 'material prosperity' as such are not to be equated with less desirable lifestyles – quite the opposite in fact (but see Taylor (1996: 93–4) for a critique of this viewpoint). Some of

the points at issue here will be taken up later; at this stage I am primarily concerned with indicating the case *against* the compatibility of green goals and democracy.

3. Green democratic arguments against ecoauthoritarianism need not necessarily be arguments against the need for a 'strong state' – which could be democratic rather than authoritarian. Some might argue that using the Rousseauan notion of democracy – 'totalitarian democracy' in Talmon's (1952) rendering – the strong government needed to enforce necessarily harsh measures which are in the public interest can be combined with such enforcement being made more effective by being given the greater legitimacy that democracy provides. However, liberal democrats would argue that such a state is really in effect authoritarian (Holden, 1993: 153–61). It can be argued, though, that a genuine, liberal, democracy can provide the necessary strong government – where, and because, there is the necessary public support. In any case, as will be argued later, the important point about legitimacy applies more authentically to genuine, liberal, democracy.
4. The inclusion of 'other species' is a typical feature of green democratic theory. Clearly, however, it is controversial; it raises issues which are tangential to our concerns, and will not be a feature of the notion of democracy to be developed in this book. The issue of future generations is, however, a different matter and will be commented upon in Chapter 3.
5. For recent discussions of the concepts of sustainability and sustainable development – some of which are pessimistic about their clarity or utility – see the contributions to Barry and Wissenberg (2001).
6. For varying perspectives on the relationship between sustainable development and liberal democracy, see, again, Barry and Wissenberg (2001).
7. It is widely held that the end of the cold war saw a 'triumph' of liberal democracy, a view especially associated with Fukuyama (1989, 1992). A basic contention here is that liberal democracy is the best form of government, and this involves the assertion that it is the best form of democracy. The grounds for such an assertion overlap with traditional liberal democratic arguments for maintaining that liberal democracy is the *only* form of democracy (see, for example, Holden, 1993). Despite some potential for confusion here, the conventional liberal democratic view that liberal democracy is the only valid form of democracy has been widely seen as vindicated. But see also the points made about the different meanings of 'liberal democracy' in the next paragraph in the main text below.
8. There can be further confusion here since green thinkers who *do* challenge the dominant view have an account of the nature of liberal democracy that would not be accepted by liberal democrats. Essentially, this is a recurrence of the cold war ideological stand-off over the concept and practice of democracy between Western liberal democrats and Marxist-

Leninists: 'Ecological Marxists are generally committed to democracy . . . . That democracy is not liberal: "to rely on the liberal democratic state in which 'democracy' has merely a procedural or formal meaning won't work"' (Dryzek, 1996: 114; quoting O'Connor, 1990: 3).

9. This is not to say that actual political systems necessarily take one or the other of these forms. Rather, these forms are something in the nature of models or ideal types, deriving from differing interpretations of the nature of actual systems, or from rival theoretical accounts of the proper nature of democracy, or from some combination of both.
10. Doubt is cast on the validity of this account of decision-making by the people, but there tends to be an oscillation between two positions. On the one hand it is contended that decisions which take this form are defective decisions by the people, but on the other hand it is maintained that they are not really decisions by the people at all. In the latter case, properly speaking, the criticism is not that a 'liberal democracy' is an inferior form of democracy but that it is an inferior system (in part) because it is not a democracy at all.
11. There can in fact be some confusion and ambivalence here since such criticisms frequently purport or appear to be wholesale critiques of liberal democracy in the wider, generic, sense. However, closer analysis reveals it to be that which is denoted by the narrower sense that is the object of criticism. This may be because of the contention that in a capitalist system it is only (so-called) liberal democracy in the narrower sense that can exist. Macpherson (1977) illustrates this point (see Holden, 1993: 148–50).
12. And, it is often argued, in the case of liberal democratic states that the fear of electoral unpopularity will ensure that governments will not deviate from carrying out this task.
13. As Barry points out in a footnote (75), this ties in with the debate about sustainable development 'which, at root, can be seen as an attempt to reconcile the interests of consumers in economic growth with their interests as concerned "green citizens" in ecological sustainability'.
14. One of the points made in Chapter 4 is that the very idea of global democracy tends to be opposed to, and to exclude, (global capitalist) processes narrowly focused on economic interests. By contrast, the traditional concept of democracy is rooted in the state, the political processes within which necessarily encompass the operations of economic interest groups.
15. See, for example, Held (1995: ch. 2). See, also the section on 'citizenship theory' in Chapter 2 of Holden (1993).
16. See, for example, Heater (1990: 229–44), Archibugi (1995: 134–5), Thompson (1998), Held (2000: 29–30). See, also, Hutchings (1996).
17. There is a large literature, but see Gutmann and Thompson (1996) for a comprehensive discussion. For a more succinct introduction, see Leet (1998).
18. The point about democracy and the ability of ordinary people to look

beyond their short-term interests and to concern themselves with long-term environmental benefits, is discussed in some depth, as it applies to the problem of global warming, in Chapter 3.

19. Referring to Miller (1992) and Dryzek (1987), Eckersley goes on to suggest that (a) deliberative democracy is best suited to small-scale communities, but that (b) small decision-making units are 'unlikely to achieve the levels of co-operation and co-ordination that are required to solve complex transboundary problems beyond the local level' (217). Indeed, it 'is uncertain how far [deliberative democracy] can be generalized for society (and international society) as a whole' (217). It is worth noting here that Fishkin (1991) has some important ideas about applying deliberative democracy on a large scale, and Dryzek (1995) develops some complementary ideas specifically with regard to environmental issues. And I shall, of course, be taking up the whole issue of the 'international application' of democracy in Chapter 4 below.
20. I am here concerned specifically with deliberative democracy's focus being on the public good. But for Barry an important and related merit is that only a deliberative process 'will reflect the range of human interests and values in respect to social-environmental relations' (216). This point has particular relevance for global democracy, where, clearly, there is an especially wide range of human interests and values to be catered for.
21. It does seem that this view underestimates the importance of the structuring presence of government in enabling the kind of cooperative individual behaviour necessary to avoid the prisoner's dilemma. Nonetheless, according to an important line of thought the prisoner's dilemma can be avoided through voluntary cooperation where there are repeated interactions among many agents – in contrast to the single set of interactions among just two persons posited in the original prisoner's dilemma (Taylor, 1987; Ostrom, 1990). There are important resonances here for the discussion of global civil society, governance (as distinct from government) and stateless global democracy in Chapter 4.
22. In liberal democracy, moreover, the area of economic activity is typically to an important extent excluded from governmental regulation. This is associated with, but also reinforces, the basic focus on economic interests; but it also means that even where there is a perception of a conflicting environmental public good, governmental action is unavailable in this area to solve prisoner's dilemma problems.
23. This argument is not compatible with the previous one, where 'raw' individual preferences reflected rational (narrow) self-interest. But common to both is the idea that the raw preferences are inauthentic in that they do not express individuals' true interests and what they would want in an 'uncontaminated' situation.
24. A 'predilection for direct democracy' usually involves an admiration for ancient Athenian direct democracy together with an acknowledgement

that some departure from this is necessary in the modern world of large-scale and complex government. It is this acknowledgement that distinguishes participatory from direct democracy. It is, though, increasingly being argued that modern information technology now makes some form of direct democracy feasible (see, for example, Budge, 1996) and the distinction between participatory and direct democracy can sometimes become blurred. At the same time, however, this sharpens the distinction between forms of participatory democratic theory which see direct democracy as an ideal and one which should be realized where that is possible after all, and forms which see merit in the differences between participatory and 'pure' direct democracy.

25. Both anarchists and Marxists have held this view, although to an extent it involves a confusion between the notions of individual and popular self-government (Holden, 1993: 151–2); see also Pateman (1985: ch. 7) for an illuminating discussion of anarchism and democracy.

CHAPTER 2

Scientific Expertise versus Popular Opinion

The Introduction identified the two broad types of reason for maintaining that democratic decision-making is unfitted for, or incapable of, dealing with the global warming problem. One was that democracy is inapplicable on a global-scale and the other was that democratic decision-making would in any event be especially unsuitable. I shall take up the global-scale issue in Chapter 4, and in the present chapter and the next I shall consider the charge that democracy *per se* is especially unsuited for coping with global warming.

This charge, broadly speaking, has two main aspects. One concerns the knowledge of the general public or the mass of the people, and the other their capacity to focus on their *long*-term interests. In this chapter I shall consider the first of these. The fundamental issue here was touched on in Chapter 1, and concerns the extent to which the global warming problem is a matter of science. Questions about the existence and nature of global warming, and possible remedies for it, would seem to be pre-eminently questions for scientific experts. The science is quite complex and it is maintained that the average person simply does not have the relevant scientific knowledge and understanding. Scientists, and not the mass of the people, are therefore those who should decide what, if anything, needs to be done. Democratic decision-making would be fundamentally unsuitable.

In this chapter I shall assess arguments such as these. Besides taking up some general issues concerning knowledge and democratic decision-making, this will involve consideration of the role and nature of science and the particular character of the global warming problem.

Guardianship

Underlying the contention that decisions about global warming are properly the concern of scientists rather than the mass of the people is Plato's critical distinction between knowledge and opinion. It was 'the dependence of government upon opinion that was the object of classical critiques of Athenian democracy by Plato in *The Republic*, on the grounds that knowledge (*epistémé*), not opinion (*doxa*), should steer the ship of state' (Weale,1999: 14). In today's world science is frequently seen as providing *epistémé*.

Stemming from Plato, then, the central traditional arguments against democracy derive from the idea of 'guardianship'. As Dahl puts it:

> A perennial alternative to democracy is government by guardians. . . . Ordinary people, these critics insist, are clearly not qualified to govern themselves. The assumption by democrats that ordinary people are qualified, they say, ought to be replaced by the opposing proposition that rulership should be entrusted to a minority of persons who are specially qualified to govern by reason of their superior knowledge and virtue.
>
> Most beautifully and enduringly presented by Plato in *The Republic*, the idea of guardianship has exerted a powerful pull throughout human history. (Dahl, 1999: 52)

Down the ages, then, the central criticism of democracy has been that as government is a matter for those with knowledge and virtue the ordinary people are not qualified to rule. As Dahl says (1999: 65), '[m]uch of the persuasiveness of the idea [of guardianship] stems from its negative view of the moral and intellectual competence of ordinary people'.

We have already seen that the guardianship argument was central to 'ecoauthoritarianism', with Ophul's 'justification of his anti-democratic stance [being] basically the traditional argument of "the ship of state" requiring the best pilots, and the dangers of "rule by the ignorant" when faced with such a complex and complicated issue as social-environmental dilemmas' (Barry, 1999: 195). And, of course, the global warming problem amounts to, or poses, a – if not the – major social-environmental dilemma of our time. We have here, then, the grounds for key arguments against the involvement of ordinary people in policy-making concerning global warming.

Initially I shall focus on knowledge regarding the phenomenon of global warming itself, rather than on knowledge relating to the 'social-environmental dilemmas' it poses. The former raises issues that deserve some separate consideration, especially regarding the nature and role of scientific knowledge. However, the full complexity of the global warming problem does, of course, involve both the nature of the phenomenon and the possible responses to it by society, or societies. Clearly, then, knowledge relating to both is necessary, and I shall take up the latter below, in this and later chapters (remembering that it includes matters such as the workings of the international system).

I am at this point, too, primarily concerned with (to use Dahl's terms) the intellectual rather than the moral competence of the ordinary people. Originally the two could not be separated since the original idea of guardianship centred on moral knowledge – knowledge of moral truths. In modern thought, however, separation is quite common. This flows from a critical distinction in modern discourse – especially salient in the case of science – between knowledge and moral evaluation, according to which it is denied that there can be 'moral knowledge'. Now, it is true that this distinction and denial are often challenged.[1] And criticisms of guardianship continue to be made that focus upon moral knowledge. But these can still be applied to knowledge of other kinds. As Harrison (1993: 160) says in his critique of Platonic guardianship arguments, 'the same points as were made [about moral knowledge] go through for other kinds of knowledge'. Since the essential points concern knowledge as such, it is to non-moral knowledge that they must be applied if this is the only kind of knowledge there is.

But even if we go along with the modern invalidation of moral 'knowledge', and the concomitant assertion of a distinction between intellectual and moral competence, we should note that when we come to the 'social-environmental dilemmas' posed by global warming there is a blurring of this distinction. There is an important dimension to the question of the competence of the ordinary people to engage with the global warming problem which cuts across, or overlaps, this distinction. It concerns the capacity of ordinary people to curb their avarice and to think and act in ways which involve sacrificing their short-term interests. To those who doubt that the people have this capacity, this is partly a matter of lack of knowledge – knowledge of the nature and importance of adverse long-term consequences of actions that further short-term interests. But it is also a matter of lack of will – the will to sacrifice short-term interests even where adverse long-term consequences are known.[2] Such lack of will can be seen as a moral defect. And even if balancing short- and long-term self-interest is not a moral matter, and hence the lack of a will to avoid long-term damage to one's own interests is not a moral defect, there are other dimensions. The long-term consequences in question may be adverse for other people instead of, or as well as, oneself. And in the case of global warming such 'other people' includes future generations. Clearly here moral questions are involved; but these will be considered later and for the moment I shall concentrate on the issue of knowledge.

What I am concerned with at this point, then, is the argument that decision-making regarding the global warming problem should be in the hands of experts – those who have knowledge of the nature and causes of global warming – rather than in the hands of the ignorant mass of the people.[3] We shall see below that the argument has various other aspects, implications and assumptions, but these undoubtedly also draw strength from its general form, which, as I have already remarked, is that of the guardianship attack on democracy. In my critical assessment, then, I shall take up the democrats' general critique of guardianship and consider its applicability to the particular argument regarding global warming.

The central idea in the general guardianship argument is the notion that only an elite has appropriate knowledge (*epistémé*) and that because of this it, and not the ignorant masses, should govern. This idea rests on the notion that there is a special set of objective truths of which members of the relevant elite have superior knowledge.

Dahl (1989: 65–6) points out that it 'is not always sufficiently noted that this kind of justification for guardianship consists of two logically independent propositions'. The first concerns the existence of 'objectively valid and validated truths' and the second is the contention that knowledge of these truths 'can be acquired only by a minority of adults, quite likely a very small minority'. And Dahl (66) goes on: 'You will notice, however, that even if the first proposition were true, the second might be false. Yet if *either* proposition is wrong, then the argument falls' (emphasis in the original). Although Dahl observes (66) that in assessments of guardianship the 'main burden of the argument . . . is usually placed on the first proposition', critiques of guardianship in fact challenge both propositions – and are strengthened by the fact that only one challenge needs to succeed.

However, this way of putting it is somewhat misleading. In fact critiques tend to interweave both kinds of challenge. Let us grant that the existence of objectively valid and validated truths concerning some matter about which decisions are to be made requires that those making such decisions should have knowledge of those truths. Now, the guardianship conclusions derived from the argument that only some can know these truths can be rebutted either by challenging the contention that only some can know these truths, or by denying the existence of such truths in the first place. But the two types of rebuttal can become interwoven to the extent that both the challenge and the denial can involve and derive from assessment of the nature of the truths claimed to be objectively valid.

Guardianship and scientific knowledge

Let me, though, start the assessment of the guardianship case with the focus initially on the claim concerning objectively valid truths. As Dahl develops it the general guardianship argument is concerned crucially with moral knowledge. And, within the terms of modern discourse, he is easily able to show that there are no 'objectively valid and validated' moral truths. Here, however, we are not concerned with – indeed we are for present purposes assuming the non-existence of – moral knowledge. Our concern is with knowledge regarding global warming, and it would seem that here at least the guardianship argument is greatly strengthened since such knowledge is held to be a matter of science. And science, according to the familiar view, does (perhaps uniquely) consist in 'objectively valid and validated truths'. In this familiar view it is science – and perhaps only science – that provides *epistémé*.

Dahl in fact does also focus on the 'truths of science' argument. He is not exclusively concerned with moral 'knowledge' and considers the contention that governing requires not moral, but instrumental, knowledge of how to obtain agreed ends such as happiness. This is essentially empirical knowledge, so that it could be said that the 'knowledge necessary to govern well could be a science like other empirical sciences' (Dahl, 1989: 67). However, this argument is specious. Dahl points out that moral judgements as well as instrumental knowledge are heavily involved in governmental decisions (he takes the example of decisions about nuclear weapons) which cannot therefore be purely a matter of science (Dahl 1989: 68–9).

In considering the global warming threat, however, the relevant knowledge is scientific, so it might seem that here the idea of the relevant knowledge consisting of objectively valid and validated truths remains untouched. But there are two main kinds of objection to this view. First, it is not in fact just science that is involved in responding to the global warming problem. As I have already said, a full consideration of the problem clearly involves the social-environmental dilemmas posed by the phenomenon of global warming and here – as in the case of nuclear weapons – moral and other, as well as scientific, judgements are involved. I shall take this up below. At this point, though, as indicated earlier, I am treating the response to global warming purely as a matter of science. This is because knowledge of the phenomenon of global warming itself is essential, and absolutely central, to understanding the problem of global warming; and since such knowledge is constituted or provided by science, the nature and function of scientific knowledge is crucially important in its own right. Here, then, it would seem that the basic idea of the relevant knowledge being objective and objectively validated remains centrally important. But this brings me on to the second form of objection to this idea, which is that there are challenges to the very notion that science provides such knowledge. These have been partly a matter of questioning the definitive role and status of science, as science is familiarly understood. But this familiar understanding of science – that it consists simply in objective knowledge – has itself come under heavy attack. I shall now turn to a consideration of these challenges.

The nature of science

Let us look first at the questioning of the definitive role of science. There are a number of lines of argument here. To begin with, even where the authoritative status of scientific knowledge and the idea that scientists therefore have especial expertise, continue to be accepted, the traditional pro-democratic argument that 'experts should be on tap and not on top' remains valid. Here the crucial importance of expertise is recognized but it is also contended that, after advice from the experts, final decisions should be made by the mass of the people. Scientific knowledge is crucially important but it does not provide sufficient knowledge. At bottom this is part of the argument, taken up below, about political decisions involving an assessment and a balancing of considerations from a range of different fields, each with its own experts. However, it also has application here, even within a *particular* field, when scientists disagree. In such cases the knowledge necessary to make the final decisions does not come only from science. Such decisions have to be made by *some* body, and there are good reasons, provided by traditional justifications of democracy,[4] for saying this body should be the people. Notoriously, disagreement among scientists is now quite common, the best known recent examples being provided by the debates about BSE and GM foods. Indeed, there is an overlap here with the next reason for questioning the definitive role of science: such disagreement now has an impact which, extending beyond simply providing an opening for popular decision-making, can impugn the credibility of science itself. As Irwin (1995: 31) says,

> when scientists . . . find themselves in public disagreement (as appears such a regular feature of policy debates), the science-centred model [of knowledge] struggles to maintain its credibility whilst more critical voices seize upon the apparent confusion in order to stress the limitations and uncertainties of scientific analysis.

And, of course, there is disagreement among scientists concerning global warming. There is now a wide consensus that human activity is causing global warming but there is some disagreement about its extent and seriousness. And there is still a minority of scientists who challenge the idea that the world is warming up, or, if it is, that this is caused by human activity. Moreover, the controversy arising from such disagreement can be quite acerbic.

Disagreement among scientists, then, besides simply highlighting the role of the non-expert public, can involve, or help generate, an undermining of the very credibility of science itself. Public distrust of science, perhaps first clearly shown in relation to atomic power but more recently manifested in the BSE crisis and the controversy over GM food, has been growing. There is a tendency for doubts about whether scientists are succeeding in finding scientific truths, to spill over into doubts about whether there are such truths to be found. This tendency is underpinned by wider theoretical developments and 'the social and intellectual conditions of our time – where knowledge claims are increasingly challenged and authority is less readily accepted' (Irwin, 1995: 32). I shall comment on this further in a moment.

It is clear, then, that the notion that there is objective knowledge of the phenomenon of global warming that can only be obtained by science, is open to challenge. And, of course, this also opens to challenge the idea that it is only scientists – experts with the relevant scientific training – who know sufficient about the phenomenon to make decisions about the global warming problem.

The challenge could be seen as strengthened if, or to the extent that, the 'familiar view' of science were to be rejected. Essentially the line of argument at issue so far has been that the function of science in providing objective knowledge is insufficient or inadequately performed. But, as was indicated just now, this line can overlap with another that questions that alleged function itself. The familiar view that science is simply, or at all, concerned with providing objective knowledge has in fact been increasingly challenged.

One type of argument here is that science, far from providing knowledge which can be used to help overcome environmental problems, is itself one of the causes of such problems. Rather than science leading us out of current environmental crises, it may be its 'very rationality which creates an exploitative and short-sighted approach to the natural world' (Irwin, 1995: 32). According to this view it is science that makes possible, and provides the driving ethos behind, the processes of industrialization and modernization which are causing the emissions responsible for global warming. This kind of view ties in with, and is often a part of, those central strands of green theory which see industrial society as the cause of our current ecological crises (Dobson, 2000: 27). One response here 'is to dismiss the application of science in this area and to turn to more romantic (or obscurantist) alternatives' (Irwin, 1995: 32), although it may be that this is more of a challenge to the use to which scientific knowledge is put rather than to the notion that science provides knowledge. Be that as it may, the central idea that the function of science is to provide secure, reliable objective knowledge of the environment is undermined.

A version of this kind of view is part of Ulrich Beck's argument concerning the 'risk society' (Beck, 1992, 1995). In today's society science can no longer be relied upon to avert risks. This is partly a matter of advancing scientific understanding of the world and of the limits of science in the face of risks, but also a matter of the changing nature of the – especially environmental – risks. As Axtmann puts it, the 'development of nuclear, chemical, genetic and military technologies has resulted in unprecedented ecological hazards and risks to the well-being and even survival of humankind'. And '[t]hese new risks differ in many ways from the "old" risks', for example by being 'incalculable' and 'limited neither in space nor time' (Axtmann, 1996: 110) – both of which qualities are notable features of the global warming threat. An important aspect of incalculability is that 'risks can only be tested and assessed after production: hence society becomes the laboratory' (Axtmann,1996: 110). Beck emphasizes the way in which this means science itself creates or magnifies risks: 'Ultimately, danger, [is] no longer subject to experimental logic . . . for nuclear power plants to be examined for safety, they must first be constructed. The application precedes the examination' (Beck, 1995: 9). A parallel point can be made about the greenhouse effect: there can only be empirical testing of hypotheses concerning the effect of emissions of carbon dioxide on global warming after those emissions have occurred.

It is upon reasons of this kind that the precautionary principle is based. I shall refer to this below, but the point being made here is that in the risk society 'science emerges as the form of understanding which has created environmental destruction'. But this is not just for the reasons already referred to. It is also because 'science is used to silence concerns about the world': '[f]ears over the environment are met with scientifically based reassurances that all is well' (Irwin, 1995: 46). Science, then, continues to purport to provide certain knowledge, but fails to do so. Irwin attributes a similar thesis to Giddens, whom he quotes as saying

> The original progenitors of modern science and philosophy believed themselves to be preparing the way for securely founded knowledge of the social and natural worlds . . . offering a sense of certitude. . . . But the reflexivity of modernity actually undermines the certainty of knowledge, even in the core domains of natural science. (Irwin, 1995: 44; quoting Giddens, 1991: 21)

Certainty has 'given way to radical doubt, reflexivity and anxiety over how each of us should live'. Science no longer provides ontological security by hiding uncertainty and risks from us and we are newly 'aware of the choices which exist' (Irwin, 1995: 44).

The failure of scientific knowledge to produce the certitude to determine decisions combines with the growing freedom modernization brings from social and institutional constraints to give a new 'social agency' to citizens to shape their lives. And, significantly, the 'environmental movement represents an excellent example of this new kind of "social agency"' (Irwin, 1995: 45). The kind of challenge to scientific guardianship arguments that can be mobilized here is developed more explicitly by Axtmann when he points to the 'questions of democratic accountability and the desirability of the rule of experts' that are raised. And he goes on:

> It is, of course, a key characteristic of the 'new' risks that they are perceived as not to be amenable to technocratic solution. On the contrary, they are often seen as caused by 'the rule of experts' . . .
>
> It has been around the rejection of expertocracy that many of the new social movements[5] have formed in the last two or three decades. (Axtmann, 1996: 113–14)

The tie up between pro-democratic arguments and the lack of scientific certainty can also be conceptualized in terms of the 'precautionary principle', which is central in environmental thought. As Eckersley (1996: 231–2) points out, 'there is already a well-established environmental decision rule – the precautionary principle – that has been designed to deal with the scientific complexity and uncertainty associated with many environmental problems'. And she quotes the words of the 1992 Earth Summit's Rio Declaration: 'Where there are threats of serious or irreversible damage, lack of full scientific certainty should not be used as a reason for postponing cost-effective measures to prevent environmental degradation.' As Barry (1999: 225–6) puts it,

> It is precisely because environmental problems are disputed (for example, global warming) within the scientific community that the precautionary principle holds that decisions ought to be made in advance of scientific proof (consensus within the scientific community), which may not be forthcoming anyway.

Paehlke (1996: 32–3) makes a similar point: 'Scientific uncertainty and ambiguity is, unfortunately, the norm in matters environmental . . . environmental policy interventions can and should precede scientific consensus .'

This kind of argument clearly has anti-guardianship, pro-democratic implications. Barry (1999: 226) spells it out in this way:

> What the application of [the precautionary] principle indicates is a challenge to the accepted relationship between science and policy-making. . . . It . . . shifts environmental decision-making away from technical or expert determination based on known 'facts', and towards making public judgements in the face of uncertainty and controversy.

There is also another, and very important, kind of challenge to the idea of science as the provider of knowledge. This denies that objective knowledge can be provided only by science, because, it is maintained, science provides no such knowledge.[6] The line of thought here ties in with postmodernism and the whole attack on the idea of objective knowledge, but is specifically concerned with the claim of science to be providing such knowledge. (And, it should be noted, such a claim is, indeed, made by those who specifically see science as uniquely immune from the general attack on objective knowledge.) This is the current of thought that stemmed initially from Kuhn's hugely influential *The Structure of Scientific Revolutions* (1962, 1970), and was taken further by Feyerabend (1975, 1978, 1987). It undermined the orthodox, 'enlightenment' or 'received view' of science as providing an objective methodology for generating empirically verifiable, and therefore objectively true, propositions[7] about the natural world.[8]

Irwin conveniently refers to analyses of this kind as the sociology of scientific knowledge (SSK). He contrasts the 'SSK analysis' with the 'enlightenment' or 'orthodox view of "science" as a form of knowledge [that] is value-free and objective, only its application [being] subject to social selection' (Irwin, 1995: 47). In the SSK perspective 'the development and social construction of scientific "facts" became a legitimate object of study' and science was no longer seen as 'a storehouse of "facts" which different social groups can plunder – nor is it a prescribed "method" for the acquisition of "objective knowledge"'. Science, then, 'emerges as a very human and – by necessity – constrained enterprise, even if its findings are subsequently presented as canonical' (Irwin, 1995: 48–9). An important view within this perspective is that the 'social construction of scientific "facts" ' flows from and reflects various social interests, so that science 'becomes a weapon used to further economic and political interests in a somewhat covert manner'. Rather than being concerned with the disinterested pursuit of, and with providing, objective knowledge, 'science essentially becomes "politics by other means" ' (Irwin, 1995: 49). Many would say this is clearly to be seen in the case of global warming where there is considerable disagreement about the nature and seriousness – and even the existence – of the phenomenon and 'scientific findings' are mobilized[9] to further social interests,[10] such as those of oil-producing states and companies. From the perspective of SSK, then, it is possible 'that science is based upon sets of assumptions about the "external world" which are social in their origination'; and '[r]ather than being the inevitable product of human enquiry . . . the science we get will reflect the social priorities and audience constructions of its sponsors' (Irwin, 1995: 51).

Science, then, according to this analysis, does not provide us with objective knowledge. But there does not necessarily have to be acceptance of this analysis – which is indeed the subject of ongoing controversy[11] – for there to be important anti-guardianship implications. The crucial point is that serious doubts *are* raised about whether science provides objective knowledge; and there has to be a way of deciding in the face of such doubts. Science itself cannot provide the authoritative answer since it is precisely the knowledge claims of science that are at issue. And just as in the case raised earlier of disagreement among scientists, this opens the door to the traditional arguments of democratic theory in favour of the people deciding. To put the contention another way, where the pronouncements of science are not accepted as definitive then decisions have to be made about these pronouncements. And it can be maintained that such decisions should be made by the people.[12] More generally, it can be said that there is mutual reinforcement between the present and the earlier arguments, outlined above, against the idea that scientific knowledge can or should determine decisions. We have already seen that these earlier arguments clearly apply in the case of global warming and there can only be mutual reinforcement between this and the application of the effects of the SSK analysis. Science is not the unchallenged provider of definitive knowledge about the phenomenon of global warming and, accordingly, democratic decisions relating to the phenomenon are not to be ruled out.

Not just a matter of science

In any case, as made clear previously, anti-guardianship arguments are not only of the above kind. What is at issue is not simply the phenomenon of global warming itself, but also the problem to which it gives rise. It is true that the former is central to the latter, but there is more to the problem than the phenomenon itself, for instance issues concerning the social and political structures and forces involved in any attempt to combat global warming. Knowledge of these matters is also necessary for decision-making; and it is certainly arguable that it is not science that provides such knowledge.[13] (The issues, in fact, are too large to be taken up here, and I shall merely observe acceptance of such an argument to be at least as prevalent as its rejection.) In other words, there are important anti-guardianship arguments in addition to, and apart from, the doubts about scientific knowledge.

But this is not really what is most important here. The paramount consideration is that it is not simply knowledge that is at issue. Or, rather, what is at issue is not simply knowledge of the immediate matter in hand. In effect there are two points here, though they are importantly interrelated. The first is that it is not just knowledge that is involved in making decisions but moral evaluation as well. The second point is that it is not simply knowledge of the matter about which the decisions are to be made that is relevant to those decisions. (They are interrelated in that moral evaluation to an important extent orders, and is ordered by, judgements concerning what additional knowledge is relevant.) And both points provide support for democratic decision-making rather than guardianship. Let me take up each in turn.

There are two aspects to the involvement of moral evaluation. First, there is the very basic point that what ends are to be pursued – what ends decisions are intended to help realize – is a matter of moral choice.[14] Important though this is in many contexts, I am, essentially, not concerned with it here since I am working within the assumption that combating global warming is the end to be pursued (see also note 14). Second, there is the argument that even where ends are agreed, and what appears to be required is simply instrumental knowledge – knowledge of the means to secure the agreed end – moral evaluation is still crucially involved. Dahl uses the example of decisions about American nuclear weapons strategies to illustrate the point. Here, he argues, despite the fact that ends are

– today we might say 'were' – agreed (survival of the United States, for example), difficult questions remain:

> The difficult questions, then, are not about ends; they concern means. But the choice of means (the argument runs) is strictly instrumental, not moral; the question is how best to achieve the ends that everyone agrees on. The knowledge required for these decisions is therefore technical, scientific, instrumental, empirical. Because this knowledge is extraordinarily complex . . . it is inherently far beyond the reach of ordinary citizens. (Dahl, 1989: 68)

But Dahl then shows (68) this argument to be fundamentally mistaken:

> To begin with, to suppose that decisions about nuclear weapons are purely instrumental and devoid of crucial and highly controversial moral questions is a profound misunderstanding. Consider some of the issues: Is nuclear war morally justifiable? If so, in what circumstances, if any, should nuclear weapons be used [and so on].

And the fact that the decision about nuclear weapons strategies 'do depend on moral judgements completely undermines the assumption that they are purely instrumental and could be made wisely on purely empirical, scientific or technical considerations'(68). Moreover, such nuclear strategy decisions are not unique. 'Decisions about crucial public policies rarely, if ever, require knowledge only of the technically most efficient means' (68). This is certainly true concerning the means to combat global warming.

The key conclusion that Dahl draws from this is that

> because both moral understanding and instrumental knowledge are always necessary for policy judgements, neither alone can ever be sufficient. It is precisely here that any argument for rule by a purely technocratic elite must fail. As in the case of nuclear forces, technocrats are no more qualified than others to make the essential moral judgements. (69)

Indeed, Dahl adds that technocrats may be *less* qualified to make such judgements. And this brings me to the point – the second of the two referred to above – about knowledge not being restricted to the immediate matter in hand. As Dahl puts it, 'the specialization required in order to acquire a high degree of expert knowledge is today inherently limiting: one becomes a specialist in *something*, that is, in *one* thing, and by necessity remains ignorant of others' (69; emphasis in the original). This is, of course, an important theme in democratic thought. What is required is a general spread of knowledge in order to be able to relate matters to one another. This is important in part because it gives more complete knowledge. For example, informed decision-making about nuclear weapons strategies requires knowledge of missile technology and nuclear physics to be complemented by knowledge of – among other things – the motivations of power holders. But perhaps of even greater importance is the diversity of knowledge required. The need to integrate decisions in different policy areas involves the appraisal of interconnections and the setting of priorities: decisions about resources to be devoted to nuclear weapons, for instance, involve judgements relating defence to other areas of state activity. And this all requires knowledge of a wide range of matters.

The argument is that it is the ordinary people rather than technocrats or experts who are going to have this wider knowledge. There are really two types of argument here. There is first the contention that it is the ordinary person, precisely because he is a non-specialist and because his life is about coping with the vagaries of a wide range of experiences, who has the range of information inputs. And this, added to the traditional democratic arguments (to be referred to later) for his knowledgeability, implies that it is the ordinary person who has the necessary breadth of knowledge. Second, there is Aristotle's argument that knowledge is scattered throughout the people as a whole; and the range of persons this brings in must imply a range of knowledge: 'although each individual may be deficient in the qualities necessary for political decision-making, the people collectively are not deficient in this way: they are, indeed, better endowed than any experts, since the *combined* qualities of all the individuals add up to a far from deficient totality' (Holden, 1993: 195–6; emphasis in the original).

But it is not – or it is not just – that democratic decision-making benefits from the range of knowledge that is located among the people. Rather it is – or it is also – that the decision-making process itself operationalizes and integrates the knowledge. (There is also the further argument of participatory democrats that genuinely democratic decision-making generates knowledge.) Barry (1999: 215) applies this argument in the context of environmental issues:

> Green arguments for democracy can be said to rest partly upon the integrative function of democracy. This refers to the manner in which democratic decision-making allows the (always provisional) determination of the social-environmental metabolism to be affected by arguments drawn from various sources of knowledge, perspectives and groups. . . . Only an open-ended communicative process such as democracy can call forth and possibly integrate the various forms of knowledge that an ecological rational metabolism will require to command widespread support.

In the mobilization and integration of the relevant range of knowledge, then, democracy, far from being deficient, is superior to guardianship. And this even – or perhaps especially – applies to what can be seen as 'scientific' or 'technocratic' issues. We should note that Barry's statement, quoted above, also points to something else. This is the overlap of the knowledge argument with other important lines of pro-democratic argument, in this case the need to command support for policy decisions. Such support is especially necessary where decisions involve difficulties and/or risk unpopularity. This, of course, is often true of decisions relating to environmental matters, especially with regard to combating global warming. I shall say something more below about the linkage with other pro-democratic arguments, and in particular the need to command popular support. But for the moment there is a complication concerning the integrated knowledge argument to consider.

An important reason for holding wider and integrated knowledge to be necessary is that what is required is knowledge of the good of the community.[15] On this reasoning, more than restricted, specialist, knowledge is necessary since the communal good embraces diverse matters. And integration is necessary because that which is good for the community is that which balances and integrates its diverse parts and their particular goods into a coherent whole. The devising of policies good for the whole not only requires knowledge of the diverse parts but also its integration into knowledge of the integrated whole, and what would benefit it. In this way the case for democracy is strengthened. The argument that wider and integrated knowledge is necessary is given an additional dimension, and – as already argued – it is democracy that best provides and mobilizes such knowledge.

But there is now a complication in the case for democracy as I shall deploy it. The 'common good account' just indicated is important in democratic theory.[16] An important reason for this is the traditional tight conceptual connection between democracy and the nation state. I shall be discussing this connection later (Chapter 4), but the essential point is that democracy has heretofore been conceived as a form of government of a community – in modern times the community that is the nation state. Democracy, then, necessarily involved making decisions about what the community should do, i.e. decisions about what is good for the community. However, one of the central thrusts of the argument in Chapter 4 about global democracy concerns the dissociation of democracy from the nation state. And this would seem to rebound on the argument about a need for wider, integrated knowledge being best met by democratic decision-making, since presumably the uncoupling from community knocks out a key reason for requiring such knowledge in the first place.

This raises some pivotal issues, some aspects of which are taken up in Chapter 4. But there are two points to be made here. First, a disconnection from community would not remove all reasons for wide and integrated knowledge being desirable. In the abstract we can say that in deciding on any set of actions there still has to be a choice of goals, a judgement of means, an assessment of advantages and disadvantages and a settlement of priorities; and all this requires a wide range of knowledge. True, there would not be the shaping and integration involved in, and required by, knowledge of a community's good. Nonetheless, some structuring to relate knowledge of proposed actions to other aspects of life is still necessary. More concretely it can be said that (in addition to knowledge of the phenomenon) contemplation of decisions regarding action on global warming requires judgements about, and hence knowledge of, many matters, ranging from the nature of – and the willingness of people to adopt – required changes of lifestyle to the dynamics of the international system. If the absence of a need to comprehend a community's good meant the required interconnections between judgements and knowledge would be less rigorous, and constrained, there would still have to be interconnections. (And, of course, the underlying argument remains that it is the mass of the people who have this interconnected knowledge.[17])

But now I come to my second point, which is that in the case of global warming it can be questioned whether there is in fact this lack of a need to comprehend a community's good. Again, this matter is taken up in Chapter 4, but the basic argument is that in this case there is a communal good. Arguably there is (at least an emerging) global community, which has its own common good. Moreover, it can be further argued that in a mutually reinforcing process global warming and global democracy are helping to create, or to make more salient, this global community. Action to combat global warming is for the good of the whole world, and the existence of such a clear global common good can itself help forge a global community. Moreover, the need for action may generate global structures and processes, which could be (at least potentially) constitutive of such a community. And the growth of global democracy would help promote, and would itself be promoted by, these developments.

In short, in respect of global warming there is in any event a case for democratic decision-making in terms of the need for a range and integration of knowledge. But it is a case that is strengthened by arguing that the knowledge required is of a world communal good.

Further anti-guardianship arguments

The particular arguments I have been considering are, of course, supplemented, or complemented, by the traditional arguments of democratic theory regarding the common man's capacity for knowledge.

To begin with there is the basic anti-guardianship contention that the common man *is* knowledgeable. 'True opinion on political and moral matters is the privilege of the common man. Accordingly power in a community should reside with him: and this it does only in a democracy. Hence the superiority of democracy' (Wollheim, 1958: 240). It can be objected that this is simply to assert, rather than to demonstrate the validity of, that which the guardianship argument denies. However, there are a number of points to be made. First, there is a wealth of argumentation and evidence – in connection with studies of voting behaviour for example – to support the contention that the common man has a degree of knowledge (see, for example, Holden, 1993: ch. 2, section 4). True, this would only establish that the common man has a certain amount of knowledge and not that he has the special knowledge of 'those who might be specially trained or otherwise specially qualified to rule' (Holden, 1993: 196). But here I come to the second point: the contention concerning the knowledge of the common man should be considered in conjunction with arguments already looked at, denying that there is only a special kind of knowledge (specifically, scientific knowledge) which is relevant and which is beyond the ordinary man's comprehension. Third, it can be argued – as participatory theorists do (Holden, 1993: ch. 3, section 2) – that engagement in the democratic process itself crucially increases participants' knowledge. This is especially relevant in countering the argument that there are complex matters which, being beyond ordinary people's grasp, are unsuitable for determination by them. Such considerations justify the use of referenda on matters as complex as whether the United Kingdom should join the euro; and similar considerations apply in the case of decisions relating to global warming. This kind of argument overlaps into another, which is equally important. This relates to the SSK analysis of science and the general postmodern relativist view of 'knowledge'. It also relates to the Popperian view of science. The argument here is that democracy is the best way of providing knowledge,[18] at least in the sense of being the best way of adjudicating rival knowledge claims. More precisely, the reference is primarily to deliberative democracy, but it is arguable that in some sense deliberative democracy is the 'ideal type' of democracy and/or that some degree of rational argumentation is an inherent property of democracy *per se*. From this perspective knowledge 'is construed fallibilistically, that is, it is never seen as final and conclusive but always open to challenge and revision in the light of new evidence and arguments, [hence] unconstrained rational argumentation seems the most appropriate forum for adjudicating rival claims' (Cooke, 2000: 955).

The fourth point is of especial importance. It concerns what is perhaps the main argument in favour of the idea that the opinion of the common man should prevail. This is the argument that whatever else the ordinary person does or does not know he at least knows his own interests better than anyone else.

It is true that there are important possible objections to, and complications regarding, this argument. These include general guardianship denials that ordinary people are the best judges of their own interests. But here guardianship arguments are at their least convincing. It is less plausible to argue that a person does not know what is in his own interest than that he lacks knowledge of matters less directly related to himself. Moreover, the guardianship arguments are dangerous. The idea that a person is the best judge of his own interests overlaps into the crucial contention that he is the best protector of that interest: guardianship can let in, or excuse, all sorts of tyrannical harming of people's interests. I shall take this up in a moment. There are also more particular and perhaps problematic applications of the general guardianship arguments here, of especial relevance in the context of the global warming problem. These mainly concern issues regarding long-term versus short-term interests. These have already been raised and will be further discussed in the next chapter. I shall leave this matter on one side for the moment; but there is also what is to an extent a parallel set of concerns regarding the relationships between the interests of particular individuals and those of other individuals and, indeed, of collectivities. There are important and complex issues here. Some of these will be taken up later, but at this point we can at least say that it remains a crucial contention of democratic theory that individuals are the best judges of what is in the common interest.[19] Despite important objections, then, to the simple identification of individual interests with the common interest, there is still

> an important core to the argument from individuals being the best judges of their own interests that remains valid. This is simply the notion that whatever else goes into the making of [common interest] decisions, in the end such decisions are only justifiable if the mass of the people judge their experiences of them to be acceptable. (Holden, 1993: 196)

There is an illuminating 'shoe-pinching' analogy which sums up the pro-democratic arguments about people being the best judges of their own interests:

> This type of argument, developed from an argument of Aristotle's, is stated in terms of a 'shoe-pinching' analogy. It is found in Lindsay (1943) . . . only the wearer (the people) knows where the shoe pinches (the effect of governmental policies[20]). . . . The expert could be said to propose and the mass dispose: the shoemaker [who has the necessary expertise] makes the shoes and the people decide whether to accept them or call for others, and perhaps for other shoemakers, according to whether they pinch or not. (Holden, 1993: 197)

Overall, then, we can say that, despite the apparent force of the guardianship case in relation to the issue of knowledge and action on global warming, the case for democracy is stronger. Moreover, traditional democratic theory's anti-guardianship knowledge arguments overlap into two others, which can further strengthen the democratic case.

The first of these is constituted by the 'all-affected principle', which in effect holds that governmental actions[21] should be subject to the control of those whose interests they would affect. It is not just because it is the wearer who knows where it pinches that it should be he who decides whether to accept the shoe; clearly it is also because it is his foot that is being pinched. At issue is the democratic principle that people's wishes should prevail, which, given that individuals are the best judges of what their interests are, here translates as meaning that it is the wishes of those whose interests are affected whose wishes should prevail. The usual idea is that the reference here is simply to 'the people' as traditionally understood. However, different ideas will be crucial to the argument of Chapter 4, where I shall further discuss the all-affected principle.

The second additional anti-guardianship argument (to which some reference has already been made) provides a powerful reason for control being in the hands of all whose interests are affected. Dahl (1999: 76) puts it this way:

> It is then highly doubtful that the guardians would possess the *knowledge*, whether moral, instrumental or practical, they would need to justify their entitlement to rule. But even superior knowledge would not be enough. Could we trust our putative guardians to *seek* the general good rather than merely their own? [Emphasis in the original.]

And he backs this up with Lord Acton's famous assertion that 'power tends to corrupt, and absolute power corrupts absolutely', and a quote from John Stuart Mill:

> The rights and interests of every or any person are only secure from being disregarded when the person interested is himself able, and habitually disposed, to stand up for them. . . . Human beings are only secure from evil at the hands of others in proportion as they have the power of being, and are, self-*protecting.* (Dahl, 1989: 76, quoting Mill [1861] 1958: 43)

Now, in the case of global warming, as I have already mentioned (see notes 3, 20 and 21), and as I shall discuss in Chapter 4, it is not essentially 'a government' that would be making decisions. (In fact the fear that it would be corrupted by its immense power is, precisely, a crucial argument against the idea of having a world *government*.[22]) Hence it is unclear whether, or in what way, there would be a definite group of people – 'putative guardians' – who would have their own interests, or would have the power to corrupt them into the pursuit of any such interests.

The argument remains relevant, however, for three reasons. First, governments – 'national governments' – would still have some involvement in the decision-making process. The traditional argument would still have application here, even if lack of overall control meant reduced power and therefore reduced scope for 'power corruption'.[23] Second, even though under anticipated forms of global governance power would be diffused, there would still be relatively important sites of power other than erstwhile national governments: democratization could be important here for (among other things) countering 'power corruption'.

The third reason is that even if, or to the extent that, power corruption was not really a problem posed by the personnel of governments and governance, there would still be organizations with worrying degrees of power or influence. There are significant actors on the international scene who have no formal power but who are nonetheless powerful – transnational companies for example. True, their current power to a significant extent results precisely from the relative lack of formal governmental power. (Within a state non-governmental actors are to an important extent subject to government control; and the idea is that democratic government would ensure that such control would be genuine, and exercised in the public interest, rather than being corrupted by the private interests allegedly being regulated.) More developed systems of global governance would modify this, but not perhaps to any crucial extent – and we are here precisely considering a scenario in which personnel of governance have no great power. Such organizations may well 'abuse' their power or influence, and need to be restrained by democratic control. The important point here is that the 'Acton argument' still applies since there is a greater risk that *non*-democratic forms of governance would be corrupted by, and act in the interests of,[24] the very organizations they purportedly controlled.

There are, of course, various additional arguments for democracy.[25] The most pertinent will be taken up later; others can be briefly mentioned here. Now, the overall project of a comprehensive consideration of the merits of global democracy is undoubtedly important, and this would involve a broad concern with the justification of democracy as such. However, I have the narrower concern of countering the application of central guardianship arguments to the issue of decision-making about the global warming problem. Although this could be regarded, or developed, as part of the overall project, that is beyond my scope here; and I am interested only in the arguments for democracy that are most directly relevant to my narrower concern.

I will not, then, comment on the 'underlying principles' of democracy (Holden, 1993: 187–91).[26] Some of the 'inherent virtues' of the democratic process (Holden, 1993: 191–5) can, though, be seen as more directly relevant to my concerns.[27] The argument that participation increases knowledge has already been mentioned. This can be broadened to include the idea of increased 'political efficacy': people would come to feel – and to be – more competent to engage with the problem of global warming. Other inherent virtues hinge on the way democracy accommodates a plurality of values and interests, and copes non-violently with disagreement. Ideas concerning the necessarily global responses to global warming, and the arrangements they require, must clearly involve a diversity of cultures and conflicts of values. Moreover, interpenetrating and exacerbating these are conflicts of interest. It should be noted, though, that perhaps the most important conflicts of interest mesh with moral disagreements which are not really connected to cultural difference. This is illustrated by the North–South disputes concerning the justness of alternative arrangements which are commented on in Chapter 4. (The propensity of democracy to facilitate the promotion of the social justice necessary to resolve these disputes, which is argued for in Chapter 4, can also be seen as a virtue of democracy which is relevant here.) In action to combat global warming, democracy – global democracy – could thus be of great value. Moreover, there is an overlap and mutual reinforcement between these virtues of democracy – and the other benefits discussed in this chapter – and that of mobilizing popular support, which is discussed in the next chapter.

Before I take this matter up, however, there is an issue that was put on one side earlier in this chapter. This concerned people's capacity to focus and act upon their long-term rather than their short-term interests. And it is this issue that I shall initially take up in the next chapter.

Notes

1. The distinction in effect disappears in postmodern discourse. Here though – as indeed in the 'post-Kuhnian sociology of scientific knowledge' (Irwin, 1995: 29) – there is not so much a denial of moral knowledge as of knowledge as such. Moral evaluation is no longer distinguished from 'knowledge', because all so-called knowledge is a manifestation of moral evaluation (which is not itself a rational activity). There are, however, other kinds of challenge to the sharp distinction between knowledge and moral evaluation which preserve both the idea of knowledge and of moral evaluation as a rational activity. I shall be commenting further on these matters later.
2. The full story is a little more complex. We have already seen that what might appear as instances of an irrational, short-sighted, preoccupation with apparent short-term interests at the expense of real long-term interests may not in fact be such. They might instead be manifestations of the collective action problem, so that the pursuit of the 'short-term interests' may be entirely rational for the individual agents. In such cases, then, a 'failure' to promote long-term interests is not any kind of defect on the part of the agents. (The failing is in the situation in which individual agents find themselves. And, as we have already seen, democracy – or democracy of a certain kind – far from being undesirable here may have the virtue of helping to overcome this failing.) Having said that, however, it is not always or necessarily the case that the collective action problem is the – or the only – culprit. There surely is an issue concerning the extent to which individuals are capable of perceiving, and/or willing to sacrifice short-term interests for long-term benefits even if, or where, the collective action problem is absent and those benefits are straightforwardly securable by relevant individual action. Further, and similarly, where a collective action problem is overcome there will remain the question of the extent to which individuals will do what it is now rational for them to do, and act to secure the long-term benefit. Thus, in the case of global warming it is true that there are crucial collective action problems arguably rendering it irrational for individual agents (be they individuals, corporations or states) unilaterally to make the material or economic sacrifices involved in reducing carbon dioxide emissions. But it is also true that there are big questions concerning the extent to which people would be willing to make such sacrifices even if, by the overcoming of the collective action problems, it would no longer be irrational for them to do so. (Of course the issue is complicated by two factors. On the one hand, there is a measure of disagreement and/or uncertainty about the alleged irrationality of not making the sacrifices, which reflects the controversies and uncertainties concerning the seriousness, likelihood and timing of global warming and

its adverse effects. On the other hand, there are the green ideas about so-called material and economic 'sacrifices' in fact being beneficial.)

3. The traditional debate between guardianship and democracy has been about whether government should be by guardians or the people. However, phrases such as 'decision-making being in the hands of, need to be used here instead of 'government' because of a key issue to be taken up later. This concerns the extent to which it would actually be 'government' that would be involved in action to combat global warming: the kind of non-state activity discussed in Chapter 4 will be classed as 'governance' rather than 'government'.
4. There are effectively three lines of argument here. First, there is the more radical democratic line (which directly challenges Dahl's second guardianship proposition) according to which ordinary people have a capacity to have knowledge, which essentially includes the necessary scientific knowledge. Second, there is the argument that the ordinary people, although they cannot have the knowledge that scientists have, should be the final judges. This is the 'jury argument' – the contention that, after hearing the rival claims of the disagreeing experts, the common people are, in the end, the best judges of the evidence. In this sense, it is they who have the relevant knowledge. Third, there is the view that irrespective of whether the people have the relevant knowledge, there are additional reasons for maintaining that they should make the final decisions. These reasons are provided by the traditional arguments for democracy – see for example Holden (1993: ch. 4, section 2). They will be taken up below, but it should be noted that there is a difference between (a) arguing (here) that there are factors which provide reasons for the people to make the decisions that are additional to the quality of the particular decisions themselves, and (b) arguing (below) that the reason the people should make the decisions is that those factors enhance the quality of those decisions.
5. I shall be commenting on the democratic function of the 'new social movements' in Chapter 4 below.
6. Whether the contention really goes this far is a matter of some dispute. However, it is often interpreted this way, and at the very least the argument (or implication) is that even if science provides knowledge of some sort, this is partial, and reflective of a perspective, rather than full and objective.
7. This is perhaps something of an oversimplification. The 'received view' can be taken here to include Popper's, as well as the 'positivist', account of scientific methodology. Whether Popper's use of the 'falsifiability principle' can be sensibly equated with accounts based on the positivist verification principle is a matter of dispute, but objective testing of hypotheses is central to both the Popperian and postivist views of science.
8. Whether, assuming the validity of the enlightenment view of science, truths about the social world can be generated in a similar way is, of course, a separate – and much debated – issue.

9. More precisely the contention is that there are either or both of two processes of persuasion involved here: (a) the selective use of scientifically established facts and (b) the use of propositions that pass for scientifically established facts but which cannot be such since all scientific 'facts' are (consciously or unconsciously) 'socially constructed'. This distinction is not always sufficiently recognized within the SSK perspective. Nor is it sufficiently recognized that (a) and (b) have very different assumptions and implications since (a) assumes, but (b) denies, that science produces objective knowledge ('scientifically established facts'). Arguably, the SSK perspective gains credibility by systematically obscuring this distinction, and surreptitiously retreating to (a) where (b) ceases to look plausible or coherent.
10. Although the term 'social interests' has a wide meaning, it is perhaps somewhat misleading here since reference is intended not only to (groups with) economic interests but also to the promotion of a cause (by groups and individuals) – the cause of combating global warming. Indeed, many might conceptualize the disagreement and conflict over global warming as being between interests and morality.
11. Scientists themselves, as distinct from 'sociologists of science', tend to reject it; but among the latter it is now something of an orthodoxy. The present writer does not accept the analysis; but then, as is argued in the main text, such acceptance is not necessarily crucial for an anti-guardianship argument.
12. The arguments of traditional democratic theory in favour of the people making the final decisions are referred to below.
13. The arguments (of traditional democratic theory) about ordinary people's capacity for knowledge in general – as distinct from the particular arguments about science I have just been looking at – are also referred to below.
14. Of course, as I have already discussed, the original Platonic guardianship argument actually made knowledge of ends central – i.e. (what is now commonly called) moral 'choice' of ends was itself a matter of knowledge. However, as I said earlier, I am adopting here the more usual modern stance of regarding ends as a matter of moral choice rather than knowledge. Moreover, because in effect the end we are concerned with is a given, we are not concerned with issues concerning the determination of ends. As explained in the Introduction, the premise of the argument of this book is that combating global warming is a morally desirable goal. It is true that this is complicated by (a) the denial of global warming by some and (b) divergences of view over the relationship of this to other goals that are often seen as morally desirable – such as the maintenance of material prosperity. There are complex issues here, but to simplify: (a) the denial of global warming does not amount to a denial that it would be morally desirable to combat it if it were occurring; and (b) the assessment of the

proper relationship between (the actually or apparently) conflictive morally desirable goals can be seen as a case of the kind of 'instrumental moral assessment' – discussed in the main text above – which is anyway involved in policy-making (the end of combating global warming is a given; the question is the balancing of the implications of pursuing this end against other – including moral – considerations). See also note 15.

15. Again, the issue of moral knowledge would seem to arise (knowledge of what is 'good' for the community). However, this can be seen as a case of 'instrumental moral assessment' rather than of a moral determination of ends (see note 14). And the essential point concerns the broader knowledge (see the main text) this requires – and the special kind of such knowledge needed where a community is central.
16. Most overtly in 'Continental' democratic theory, and in 'neo-Idealist' and 'citizenship' theory (Holden, 1993: ch. 2), which have affinities with the now important theories of deliberative democracy. It is true that more individualistic varieties of 'Anglo-American' democratic theory (Holden, 1993: ch. 2) tend to spurn such accounts of community and the communal good, but traditional democratic theory's presuppositions about democracy's linkage with national communities (see the main text above) make at least implicit references to the community's good almost impossible to avoid.
17. Two further points should also be borne in mind. First, the argument about the knowledge of the mass of the people is, of course, combined with other arguments – such as it being those who are affected by them who should have (at least) the final say on decisions – in the full case for democratic decision-making (see the main text below). Second, in relation to the global warming problem there are vital issues regarding who 'the mass of the people' may be. These issues are taken up in Chapter 4.
18. More precisely the reference should be to the best way of providing knowledge, or of making decisions in the absence of knowledge (i.e. if it is held that there can be no certain knowledge).
19. The term 'the common interest' is used here to bracket the issues concerning the communal good discussed above. Although the term can be seen as equivalent to 'the common good', this latter term in fact is normally firmly linked to – indeed is usually synonymous with – 'the communal good' or 'the good of the community'. It is, perhaps, easier to detach 'the common interest' from this tight linkage; and here the term is used to denote an interest 'all' can be said to have in common, while leaving open to what 'all' refers, and whether this interest is the realization of the good of a community.
20. As was indicated earlier (see note 3 above) in the context of global warming it is not really 'governmental' policies that are involved.
21. The same qualification regarding the reference to 'governmental' actions applies here.

22. It might, of course, be argued that this could be avoided by – and is a reason for having – a world system of government that is democratic. However, it might well be doubted that on a global scale democratic mechanisms would work effectively enough to ensure that the government was in fact democratically accountable. In any case, I rule this out for consideration since I argue that the only feasible form of global democracy is non-governmental.
23. The question of the extent to which 'national governments' would remain or have real power in a global democracy are matters to be taken up in Chapter 4. Hence attempts to assess whether they would really remain sites of 'power corruption' are perhaps premature. But in any case, if attempts to combat global warming were to continue without significant moves towards global governance then national governments and their power would remain. In these circumstances the 'Acton argument' would remain fully applicable. It may be that in such circumstances governments would be unable to tackle global warming, but in attempting or purporting to do so there could well be ample scope for 'power corruption'. Democracy – in its traditional form, i.e. within states – would remain important as a counter to this.
24. There is also the point (conceptually very different, but in practical consequences hard to separate from the point about corruption) that, try as they genuinely might, organs of governance may be *unable* to control the organizations they are supposed to regulate. Up to a point this would apply to democratic as well as to non-democratic organs; however, because such power as they would have would be legitimated, democratic organs would probably be more effective. The point about legitimation is taken up later.
25. For reasons explained in the last chapter, the reference here is to liberal democracy. But, as we also saw in the last chapter, there are different kinds, or understandings, of liberal democracy, and some of the arguments referred to below in the main text will properly apply only to some of these.
26. The relevant arguments in the 'beneficial results' category (Holden, 1993: 195–7) have already been covered in my discussion of knowledge and guardianship.
27. There are some complications here to do with the differences between traditional, state-centric, democracy and non-state global democracy. However, the significance of these is diminished to the extent that both state and global democracy involve accommodation of views within a community (the global community in the case of global democracy).

CHAPTER 3

Popular Opinion and Tough Policies

Short-term versus long-term interests

In the previous chapter questions were raised about the capacity of the mass of the people to give priority to their long-term, over their short-term, interests. These were left on one side, to be discussed in this chapter. Here, then, I need to consider the implications of these questions for the pro-democratic, anti-guardianship arguments regarding people's interests that I looked at in the last chapter. The basic question concerns the extent to which these arguments are still successful when long-term interests are distinguished from short-term interests and are specifically taken into account.

The core issue involved here concerns the extent to which people are prepared to make short-term sacrifices for the sake of long-term gains. The essential point is that climate change forces 'us to confront how much we are willing to sacrifice today for benefits[1] which will be enjoyed later in our lives or in the lives of succeeding generations' (Portney and Weynant, 1999a: 2).[2] This, of course, raises two different, although overlapping, types of question, one concerning future benefits to be enjoyed later in our own lives and the other about the benefits accruing to succeeding generations. Both will need to be considered.

Questions of the former type concern 'time preference', which 'relates to impatience about one's own future [benefit], that is, the [benefit] in the future by the person who currently forgoes consumption in the interest' of this future gain (Schelling, 1999: 100).[3] The underlying point is that, whether or not this is rational,[4] short-term interests tend to have a greater salience than long-term interests. As Kavka and Warren (1983: 28) put it: 'It is . . . a notorious fact about human nature, that people tend to *overdiscount* long-term gains and losses relative to short-term gains and losses' (emphasis in the original). Or, as ex-Vice President Al Gore puts it more pithily: 'The future whispers while the present shouts' (Gore, 1992: 170). And, of course, a common inference from this brings home the negative implications of this tendency for the role of democracy in tackling global warming: 'politicians in democratic states, who are elected for relatively short periods and who are judged by voters largely in terms of the immediate results of their actions, also have strong incentives to overdiscount the future in the policy-making process' (Kavka and Warren, 1983: 28).

According to Bentham's felicific calculus, the value of a pleasure depends not only on its intensity but also on its temporal proximity, with the most valuable pleasure being one that is now present:

> The quantum of the value of a pleasure in point of proximity has for its limit on the side of increase actual proximity. No pleasure can be nearer, no pleasure can, on the score of proximity, be more valuable, than one that is actually present. (Parekh, 1973: 115: excerpt from a Bentham manuscript)

Thus, where there are two pleasures of equal intensity, choosing that which will occur the soonest will bring about the greater happiness. Changing to a more meaningful conceptualization I shall say that where two benefits of equal intrinsic worth are due to occur at different times, that which will occur sooner will be perceived as bringing the greater happiness. In utilitarian terms, then, a benefit's impact is increased by its temporal proximity, so that its overall worth can be greater than that of one which is more distant but which is intrinsically worth more. We might speak here of a 'felicific equation' in which the overall worth of a benefit is the product of its temporal proximity and its 'intrinsic worth'. This way of putting things may not command general agreement. Nonetheless, in ranking benefits it may be impossible to avoid giving some sort of negative weight to the elapse of time before they occur, and this equation is helpful in capturing this point. There is, of course, an issue concerning what is the 'correct', or a reasonable, weight to give; and I shall take this up below.

In using the equation in any particular case different 'answers' can result from a difference in either the intrinsic worth ascribed to the benefit or the negative weight given to the time lapse before its occurrence, or both. Now, it follows from the premise of my whole argument, as set out in the Introduction, that in the case of global warming there is an answer that is *right*. And this is to the effect that the overall worth of the benefit brought by action to curb global warming is greater than that of the benefit brought by the avoidance of such action.[5] This answer flows from judgements (whose validity, it is maintained, can be defended) about the intrinsic worth of the benefit and the negative weight it is reasonable to give to the time lapse. It is contended, then, that the great intrinsic worth of the benefit brought by action, not only exceeds the intrinsic worth of the benefit brought by avoidance of action, but does so to an extent sufficient to offset the time lapse – thereby giving it greater overall worth. That is to say, the overall worth of the benefit of action to curb global warming remains higher than that of avoidance of action, even after factoring in the negative 'score' it is reasonable to give to temporal distance.

However, matters can of course come to be viewed differently, so that the greater worth is accorded to the benefit of avoiding action. The guardianship assessment is that, due to ignorance or moral weakness, this will be the view of the mass of the people. My task is to show that such an assessment is mistaken. Or, more precisely, my argument will be that with educative involvement and participation in decision-making about global warming this will not remain the view of the mass of the people, even if it is their view now.

The differing and – according to my premise – mistaken view arises because of differing judgements about the importance of temporal distance, or about the relative intrinsic worth of the benefits of action and its avoidance, or both.

Mistaken judgements about the intrinsic worth of the benefits of action can arise from ignorance about the existence or nature of those benefits; and I have already looked at guardianship arguments concerning the ignorance of the people. There can also be mistaken judgements about the worth of the benefits of avoiding action – judgements which mistakenly inflate the intrinsic worth of such avoidance. These may, however, be due to misperceptions rather than simple ignorance. I shall take this up below.

Let me first consider the issue of temporal distance, and the matter of mistaken judgements concerning its importance.

To the extent that this is an issue which concerns people who could potentially experience either the long- or the short-term benefits, it remains in fact closely tied into the issue of these benefits' relative intrinsic worth. I shall take this up in a moment. But first I need to focus on another aspect of the temporal distance issue.

This further aspect arises to the extent that the issue is not one pertaining only to people who could potentially experience either of the benefits. The point here is that the relevant time scale may need to be extended beyond the present generation. Now, one factor is the longer the time scale the greater its importance; and the more the relative overall worth of the benefits would become a matter of temporal distance as well as intrinsic worth. But this is not the only dimension of the issue here.

The extra dimension is that the time laspe I am considering, before the onset of the full benefits of curbing global warming, is such that it is only individuals not yet born who will have full experience of these benefits (the question of the extent of the time lapse will be taken up later).

In fact questions of two kinds arise here. First it may be asked by how much is the intrinsic worth of the long-term benefit to existing individuals reduced by the limited extent to which it will occur in time to be enjoyed by them.[6] Clearly, answers will turn in part on the proportion of the full benefit to be enjoyed by existing individuals. And this depends on the time scale of global warming, the extent and efficacy of efforts to combat it and the time taken for such efforts to take effect. These are all matters about which there is a considerable degree of uncertainty. This uncertainty, added to that concerning the extent and impact of the global warming phenomenon in the first place, increases the complexity of the issues I am considering. I will comment later on the complicating effects of such uncertainties; and I shall leave them aside for the time being. However – and this brings me to questions of the second kind – matters are also complicated in another way by a different consideration. This relates to the extent to which it is proper to consider the benefit to be enjoyed by individuals who do not yet exist. Here there are no questions about the reduced worth of the benefit, since I am now focusing on individuals of the future who it is assumed *will* be enjoying the full benefit. Rather, the questions now raised are these. First, how much extra negative weight should be given to the greater lapse of time that must now be involved? And, second, to what extent is it valid to extend the reference of the felicific equation beyond existing individuals, to encompass the worth of the 'long-term' benefit to those who do not yet exist and for whom the question of the comparative worth of a rival short-term benefit does not arise? Again, I shall leave such questions on one side at this point as the issue of the weight to be given to the interests of future generations is taken up below.

The long-term interests of existing people

Let me return, then, to the matter of people now living. For the moment I shall make the assumption that it is simply to them that the issue of temporal distance pertains. As already indicated, my premise is that the worth of the benefit brought by action to curb global warming is not only superior to any benefit gained from the avoidance of such action, but, further, that the degree of superiority is such as to more than offset any negative weight that it is reasonable to give to the time lapse before it occurs. More precisely, it is held to be true that for the people who will have sufficient enjoyment of it, the intrinsic worth of the long-term benefit is great enough to make it worth more overall than the competing short-term benefit, whatever the time lapse. (The term 'sufficient enjoyment' indicates that people may encounter sufficient of the benefit to experience most of its intrinsic worth even where it is not fully realized in their lifetime.)

Despite this, however, by giving the time lapse more than its proper weight it may be that it is the long-term benefit which can come to be seen as worth less overall. And the idea that this is how the mass of the people will see it is the aspect of the guardianship case with which I am now concerned.

There are in a sense two sorts of contention here. On the one hand it is maintained there is a common tendency simply to give an unduly high negative score to temporal distance in the felicific equation. Here there is a difficult issue concerning what is the 'correct', or what is a reasonable, score to give. But, as I indicated in the last chapter, part of the guardianship critique of the ordinary man is that he lacks, as it were, the 'moral competence' to ascribe a reasonable score. That is to say, it is argued that the ordinary man lacks the patience, or moral fibre, to turn aside from lesser immediate benefits and to wait instead for greater, future, benefits. On the other hand it is maintained that there is a tendency to misperceive the long- and the short-term benefits, with the intrinsic worth of the former being undervalued and that of the latter overvalued.

These two types of contention are, however, so interconnected that it may be better to speak simply of one line of argument, to the effect that the people misperceive the relative overall worth of the short- and the long-term benefits. Thus the 'moral weakness' aspect of the accusation becomes watered down to the extent that it is undervaluation of the intrinsic worth of the long-term benefit that is emphasized. The point is that the smaller the degree by which the *intrinsic* worth of the long-term benefit is seen to be superior, the less any perception of its *overall* worth as inferior must be put down to an overweighting of the time lapse. Indeed, this diminution of the intrinsic worth ascribed to the long-term benefit may flip over into perceiving it as actually worth *intrinsically less* than the short-term benefit. There might even be self-deception, or rationalization, involved here: undervaluing the intrinsic worth of the long-term benefit could obviate any 'morally damaging' implication that the time lapse was being overweighted. But then again insufficient attention may be given to the long-term benefit *because* it is temporally distant; i.e. failure to focus properly on the nature, and to appreciate the intrinsic worth, of the long-term benefit can itself result from letting the time lapse have too much weight.

The guardianship argument being assessed in this section, then, includes the charge that the mass of the people misperceive the nature, and hence the relative intrinsic worth, of the short- and long-term benefits. To an extent this raises issues concerning people's knowledge of the kind considered in the previous section. But in the anti-guardianship arguments of the previous section, which were concerned with showing that the people are capable of having the requisite knowledge to know what benefits them, the benefits were conceptualized as short-term. Or, at least, there was no contrast postulated between short- and long-term benefits. Now, however, when this contrast *is* at issue, the matter is no longer simply one of requisite knowledge. Aside from – although, as just indicated, connected with – the issue of the time lapse there is the difference that the matter is now one which centrally involves judgement and interpretation. (This is not to deny the role that judgement and interpretation must perhaps always have in knowledge and its acquisition. But unless there is a complete denial of the principles of empiricism, a workable and important distinction must for many purposes be drawn between having knowledge of, and making judgements about, something.)

What I now need to look at, then, is the matter of interpreting and judging the nature of, on the one hand, the short-term benefits of avoiding action against global warming; and, on the other hand, the long-term benefits of undertaking such action. I need to look at this so I can consider the charge that the mass of the people do, or will, misperceive the relative intrinsic worth of each of these benefits.

The essential issue in considering this charge arises from the nature of the short-term benefit, which is widely seen as consisting in the avoidance of the harsh measures held to be necessary for curbing global warming; and of course this has obvious appeal. Action to combat global warming includes measures – to reduce energy use, for example – which involve material sacrifices. Here we have a benefit – the avoidance of material sacrifices – that tends to be seen as clear and substantial and of obvious intrinsic worth. And, naturally, the overall worth of the benefit is magnified by temporal proximity. By contrast the benefit brought by attempting to curb global warming can be seen as unclear and hence of diminished intrinsic worth. And, of course, the effect of temporal distance is further to diminish its overall worth.

Possible complications

Before I turn to a consideration of this issue we need to remind ourselves of an important premise of the argument of this book, and make explicit its implications here. The premise – as explained in the Introduction – is that unchecked global warming will bring great harm, so that it is indeed the case that the intrinsic worth of the long-term benefit brought by measures to curb global warming is superior to that of the short-term benefit of avoiding such measures. Moreover, the degree of superiority is such that the time lapse does not prevent the *overall* worth of the long-term benefit being greater too. This has underpinned our characterization of these benefits and a possible disposition to see them as other than they are. The implication is that such a disposition would be mistaken. And it is important to bring out a further implication, which structures the whole character of my argument.

The point here is this. If the mass of the people do have the perception that the long-term benefit is worth less than the short-term benefit, then democracy would be an obstacle to action to combat global warming. This, indeed, is the burden of the guardianship contention. But what if such a popular perception were in fact *correct*? We would then have an argument against attempts to curb global warming instead of one against democracy. But it is exactly this possibility that is ruled out by my premise.

Now, although such an argument does not arise for us, since the possibility of such a popular perception being correct is exactly what is ruled out by our premise (as explained in the Introduction), there are some clarifications to be made.

First, it may be objected that in the previous section I did make use of the pro-democratic argument that there are good reasons for believing popular perceptions to be correct, whatever those perceptions may be – i.e. that ultimately the people are the best judges of what is correct. An important form of this argument was, of course, that the people are the best judges of what is in their interests. However, by failing to make a distinction, implicitly this argument referred to short-, rather than long-term interests; and now, of course, I am raising the different question of whether, or to what extent, the argument applies to long-term interests.

Second, there is the pro-democracy argument to the effect that it may not be possible to say what is objectively 'correct'[7]; and that (perhaps *because* of this) the people should decide what is to be done. Quite simply, then, according to this argument, if the will of the people is for the short-term benefits then this should be given effect, and there should not be action against global warming. Again, though, this is an argument excluded by my basic premise, according to which it *is* objectively correct to say that people's long-term interests in attempting to curb global warming are their true interests. If they think differently this is a failing on their part and not an argument in favour of having action against global warming blocked by democratic decision-making. And the guardianship position I have to rebut is that since the people will think differently we ought to put decision-making on global warming in the hands of a wise elite.

But this brings me to the third clarification. Under one reading of 'the paradox of democracy' the contention that the people's choice should be followed simply because it *is* the people's choice, re-emerges. In this case, though, the proposition is that the democratically chosen policies ought to be followed even if they are wrong – even if they are contrary to people's true interests.[8] This, however, is not an issue that directly concerns me here. My question is not (or not directly) the general one of whether guardianship rather than democracy is right, but the more particular one of whether guardianship rather than democracy is necessary for action on global warming. The concern we may have with the general issue of democracy versus guardianship comes from, and is a part of, the attempt to answer this more particular question. (There is, nonetheless, some overlap and reciprocity here. While general arguments against guardianship help to establish that democracy will not inhibit action on global warming so, too, the general case against guardianship is bolstered by arguments for the competence of the mass of the people in the particular instance of global warming.)

My concern, then, will be to show that the guardianship contention is wrong: that the mass of the people do have the capacity to focus on their long-term interests and would not use their democratic power to block action on global warming.

Before taking this up, however, we should note a complication that is introduced by a line of argument that 'half accepts' that the mass of the people are incapable of prioritizing their long-term interests and that they will use their democratic power to block needed action. The central notion here is that a 'self-binding' strategy is needed. This accepts that the people are indeed liable to be short-sighted and prioritize their short-term interests. But it also maintains that under the right conditions they are capable of doing otherwise – and that under these conditions they should bind themselves not to act in a short-sighted way. Constitutions are often seen in this way, and one could envisage a situation in which ordinary legislation promoting short-term interests inconsistent with a long-term strategy to combat global warming could be blocked by a constitutional provision which could only be overcome by a supermajority. Arguably, people might agree to such a provision in a referendum if it were proposed as a matter of principle, particularly if the time were right – say after a Chernobyl-type disaster – whereas they would not do so in the context of a consideration of specific policies, such as the raising of petrol prices.[9] And it is further arguable that people may come to reflect on the seriousness of the threat of global warming in a way that might incline them to agree to self-limiting constitutional provisions.

Viewed in this way, then, this is self-limitation of the people. The analogy is with an individual limiting himself. One expression of this is

> the 'Peter drunk-Peter sober' paradigm of Stephen Holmes: 'A constitution is Peter sober while the electorate is Peter drunk.' For further illustration, Holmes switches to another picture: 'Citizens need a constitution, just as Ulysses needed to be bound to his mast.' (Stein, 1998: 422, quoting Holmes, 1993: 1)

Now, this kind of point can, of course, be used to develop a *contrast* between democracy and constitutionalism (Stein, 1998) in which the constitution is seen as a necessary brake upon the power of the people (for a general discussion of constitutionalism and democracy see Elster and Slagstad, 1993). From this viewpoint, then, it would be accepted that long-term interests are indeed at risk in a democracy, and hence that a *limitation* of democracy is needed to ensure that they are promoted. However, this is to ignore the initial characterization of this restraint as *self*-limitation. Here the restraint itself is seen as an expression of the 'real' or 'authentic' will of the people – in a way that is reminiscent of Rousseau's general will – so that the people are (at least under the right conditions) more than capable of prioritizing their long-term interests. It is true that the notions of 'self'-limitation, and of the restraint as a democratic expression of the will of the people, can be questioned where a constitutional restraint has been laid down by past generations:

> If one understands democracy as the rule of the people's will . . . how can binding this will to a constitutional framework, which is the work of predecessors centuries past be justified? How can the will of [these predecessors] . . . rule the will of the actual popular sovereign? (Stein,1998: 422)

However, the presumption here that 'the people' properly refers only to the present generation can be challenged. I refer below to a 'transgenerational' or 'intergenerational' conception of the people, and it can be argued that the decisions taken by one generation pertaining to generations yet to exist are still expressions of the will of the relevant people. True, our intergenerational conception of the people crucially involves a feature that has heretofore been lacking – mechanisms for representing generations not yet existing at the time decisions are actually taken. And this means the decisions of generations existing before today cannot be seen in this way as embodying a will that includes that of present-day people. But this is to miss the point that global warming is a contemporary and future problem, and that it is the present and future generations we are concerned with. That is to say, we would only be concerned with constitutional provisions yet to be decided on that were binding on future (as well as present) generations; and these *could* be seen as decisions by an intergenerational people, if the necessary representational mechanisms were in place. In other words, from the perspective of an intergenerational people, constitutional limitations binding future generations are simply a special case of 'self-limitation' – of a people binding itself.

I revert, then, to the idea of 'self-limitation'. But whatever line I take on the interpretation of 'self-limitation', and whether or not I argue that such limitation is necessary for the people to be able to prioritize their long-term interests, the key issue remains whether in *any* circumstances the people are capable of such prioritization. In other words, I come back to my basic concern with showing that the guardianship contention is wrong and that the mass of the people are capable (whether or not only under special conditions) of acting to promote their long-term interests and would not use their democratic power to block action against global warming. And it is to this issue I shall now return.

Further issues

In fact I have more than a negative concern with simply denying that democracy will block action. I shall go beyond this to argue that democracy actually has a positive role. I shall maintain that beyond simply not preventing action on global warming, democracy will actually encourage – and may, indeed, be essential for – such action.

I need to show, then, that it is mistaken to think people will fail to focus on the long-term benefits of combating global warming, giving priority instead to what can be seen as their substantial short-term interests in avoiding material sacrifices. In fact my argument will not really turn on whether or not the mass of the people currently fail to focus on the long-term benefits. Rather, it will be more centrally concerned with the effects on the mass of the people of greater participation by them in decision-making on global warming.

A necessary condition of focusing on the long-term benefits is to have knowledge of them. Now, as already remarked, the issue of people's knowledge was the subject of the previous section, and at this point in the argument it can be taken that the people are capable of having sufficient basic *knowledge* of the long-term benefits. The issue here is primarily one about judgement and interpretation rather than knowledge. True, it can be argued that knowledge of the long-term benefits in itself involves recognition of their real – and superior – worth, and that as people gain this knowledge such recognition will come too. But such an argument is complicated by the matter of judgement and interpretation.

The point is this. It does need to be acknowledged that there are some provisos and difficulties regarding the substance, and acceptance, of the truth about the long-term benefit which must affect the prospects of its being recognized and acted upon. As already indicated, the premise of my argument precludes interpreting this as casting doubt on the core truth that the long-term benefit *is* worth more. And people's uncertainty about accepting this truth cannot be construed to imply that the people determine *whether* it is true. Nonetheless, in two crucial and overlapping respects the matter of the intrinsic worth of the long-term benefit is subject to popular determination. First, the core truth can only be 'operationalized' by being popularly accepted – people have to come to know it. Second, there is scope for judgement about, and democratic determination of, matters within the core truth. That is to say, the provisos and difficulties pose issues of judgement and interpretation, and these are properly to be settled by judgements by the people.

These provisos and difficulties centre on the degree of vagueness and uncertainty regarding the nature and timing of the benefits in question. Even leaving aside the judgements of those scientists who question its existence in the first place there is considerable uncertainty about, and controversy concerning, global warming – for instance, about the time scale for its onset and the severity of its effects. And there is a corresponding haziness about the benefits of acting to curb it. Moreover, the geographical incidence of its effects is uneven and poorly predictable; and in some areas they may even to an extent be advantageous.

The uncertainties over timing complicate the 'felicific equation'. On the one hand, it might be said that the uncertainty increases the negative weight to be given to temporal distance. Uncertainty may magnify the effect of distance. The perceived impact of events, already diminished by distance in time may be diminished further by uncertainty. On the other hand, to the extent that uncertainties of timing imply the possibility of the *earlier* incidence of the harmful effects of global warming, then the timing begins to have a positive rather than a negative weight. And, indeed, it is important to note that there is now increasing evidence of this early incidence. In fact it is increasingly being argued that we are already suffering some of the harmful effects of global warming.[10] It is possible, indeed, that as the evidence that we are already being harmed by the effects of global warming piles up, the perceived terms of the felicific equation will change, and the overall worth of the benefits of action against global warming will be boosted by temporal proximity rather than diminished by temporal distance. This is already happening to some extent, at least among policy-making elites.[11]

I shall, however, here assume that the most serious effects of global warming will only be felt at a time in the future sufficiently distant for the 'felicific equation' to be an important factor. (Whether or not this is the more scientifically credible assumption – and it may be that it is – it provides the proper, because sterner, test for the democratic rejection of the guardianship position.) This means taking very seriously the 'provisos and difficulties' standing in the way of even a knowledgeable people perceiving, and acting upon, the truth about the long-term benefits of acting to curb global warming. These are very important matters, and some would see here a decisive objection to the idea that a response to the problem of global warming should depend on democratic decision-making. This takes me back to the core difficulty, mentioned earlier, about the 'negative incentives' to which democratic politicians are subject. Democracy, it might be argued, is already making it difficult to respond to global warming: it is notorious that politicians are reluctant to advocate, for instance, energy reduction policies likely to be unpopular with their electors – and more democracy would mean greater difficulties.

There are six (not always consistent) main lines of counter-argument. First, it can be denied that policies to combat global warming are actually against people's short-term interests. Second, it can be disputed whether, or to what extent, it is true that popular opinion rejects action, harmful to short-term interests, which is necessary to combat global warming. Third, even if there is a tendency to focus on short-term interests people nonetheless *can* recognize the existence and importance of their long-term interests, and that it is engagement in a democratic process that helps them to do so. Fourth, and following on from this, it can be argued that precisely because of the tendency to focus on short-term interests, the democratic process is invaluable in countering this. That is to say, the democratic process can be seen as securing that legitimacy for, and acceptance of, policies for pursuing long-term interests which is necessary for their success. The problem of short-term sacrifices versus long-term benefits, which inevitably bedevils *any* kind of global warming policy-making process, can be most effectively dealt with by democracy. These third and fourth lines of argument add up to a powerful case for democracy – that, because of its educative and legitimizing effects, far from being unsuited for securing short-term sacrifices for the sake of long-term benefits, the democratic process may be the only policy-making process that is effective in this way. Fifth, and in relation to the previous two arguments, it can be maintained that it is after all the people who should determine just what their long-term interests are: as already argued the issues of judgement and interpretation regarding the long-term benefits of action to combat global warming should be determined by the people. Sixth, it can be argued that to the extent that short-term interests are salient in people's responses to the problem of global warming this is in important respects due to the divisiveness of the states system and the associated focus on narrow and short-term national interests. This is, indeed, a crucial dimension of the problem of responding to the threat of global warming, and one which I shall take up later. A central part of my argument will be that a remedy for the divisiveness of the states system is global democracy. But what should be noted here is that the quality of being democratic, which is enmeshed with that of being global, is crucial; and this sixth argument interlinks with the previous three.

The first of these lines of counter-argument challenges the orthodox idea of policies to combat global warming being characterized by a reduction in living standards. The orthodox idea involves conceiving living standards in the usual way as material standards of living. And the usual perception is that combating global warming must involve reductions[12] in the levels of the possession and consumption of material goods by which such standards of living are defined, since it is precisely a reduction in the production and consumption of these goods that is necessary if carbon dioxide emissions are to be reduced.[13] And indeed it is arguable that a successful response to global warming must really involve the dissolution – or at least a winding down – of the whole system of industrial capitalism upon which material standards of living depend.

The challenge to this orthodoxy can take two main forms (apart from denying that there is a problem that requires a response in the first place). First, the whole notion of conceiving standards of living in material terms can be disputed. It can be argued that the quality of life has to do with something other, or more, than material goods; and that a slowing down, or retreat from, industrial capitalism should not involve a reduction in true living standards. Reference has already been made to this theme in green political thought. Second, it can be contended that the processes of industrial capitalism need not involve damaging amounts of carbon dioxide emissions, or environmental damage generally. Indeed there can be arguments to the effect that the forces of capitalism can actually be utilized to protect the environment: for recent examples see Hawken *et al.* (1999) and Lomborg (2001).

Both of these very different forms of argument are clearly very important. However, although there will be occasion to refer to them they will not be examined in this book. There are three reasons for this. First, they take us off into spheres too large and important in their own right to be adequately considered here: on the one hand there are the important themes in green political thought just mentioned, and on the other there is a developing field of economics concerned with market solutions to environmental problems (mentioned in note 12). Second, despite their obvious importance these lines of argument command only minority support; and to be convincing a pro-democratic argument on responding to global warming should not rely on such support. This leads on to the third reason, which is that to be properly tested such an argument should eschew controversial favourable assumptions: a robust argument for democratic action on global warming must confront head on, and accept as real, difficulties concerning long-term versus short-term interests.

This brings me to my second line of argument in challenging the guardianship response to these difficulties. This accepts as genuine the conflict between short- and long-term interests, but disputes the extent to which public opinion really is against action which is harmful to short-term interests but which promotes long-term interests by combating global warming. In fact the evidence is mixed here. But in any case there may well be reason to believe that public opinion will begin to become more favourable to 'tough' policies, which further long-term interests, with the growth of both evidence about the seriousness of the threat of global warming and the concern with which this is treated by informed opinion.[14]

And this leads on to the third line of pro-democratic counter-argument, which focuses on the potential for changes in public opinion. Here it is accepted not only that there is a genuine conflict between long- and short-term interests, but also that there is a tendency for people to concentrate on the latter, even where the long-term benefit is worth more. At the same time, though, it is denied that this means people are incapable of recognizing and pursuing their long-term interests.[15] Moreover, and this is the crucial contention, it is precisely engagement in a democratic process that facilitates a recognition of the importance of promoting long-term interests, that is to say the importance of doing that which will help to obtain temporally distant but worthwhile benefits, even though this be at the expense of prejudicing conflicting short-term interests.[16] These short-term interests consist in securing benefits which already exist, or are impending. They might seem to be worth more than the long-term benefits, but recognizing them to be actually worth less is part of what is involved in the recognition of the importance of promoting long-term interests.

Educative and legitimizing functions of democracy

The argument, then, is that even if, or where, people currently concentrate on their short-term interests this does not necessarily mean that they are unable to prioritize their long-term interests, in appropriate circumstances. Essentially, such circumstances are those where engagement in the democratic process has an educative effect which changes public opinion. This can involve either top-down or bottom-up processes, or both. Central to 'top-down processes' are campaigns by political leaders and/or environmental groups, whereas 'bottom-up processes' involve positive participation by the people in decision-making and the educative effects this has.

The educative effect of leadership and campaigning is a well-worn theme in traditional 'conventional' democratic theory of the sort associated with John Stuart Mill (Holden, 1993: 68–9). It is encapsulated in the account of the role of representatives found in this type of democratic theory. (An excellent concise characterization is provided by Hanna Pitkin's summary of this – or what she calls Liberalism's – account of representation (Pitkin, 1967: 205); quoted in Holden (1993: 74); see also note 17 below.) And this theme fits, too, with modern elitist democratic theory. The key notion is that the mass of the people can be educated, and public opinion thereby changed, by elected leaders.

In fact two, overlapping, ideas are involved. First there is the stress on the function of leadership in a democracy. Here it is held that an important duty of those elected to power is to lead and educate the public to embrace policies which are beneficial but which they would not otherwise see as such, and would reject.[17] There are many instances of this idea being brought into play. For example, in Britain advocates of the policy of joining the euro urge the Labour government to exercise leadership by campaigning to stimulate support for this policy among a sceptical public. Second, there is the focus on how performance of this duty is fostered by, and meshes into, the requirement that the people make 'the basic determining decisions on important matters of public policy' (Holden, 1993: 8). It is, after all, this requirement to make – and the fact of making – policy decisions that involves the people with the policy issues. This encourages them to engage with those issues, not least by attending to the (existing and aspiring) leaders' campaigns. Campaigning environmental groups perhaps perform a similar function, though their activities might, by contrast, be viewed as a part of the bottom-up processes[18] (I shall be saying more about – in this case global – environmental groups in Chapter 4).

When we turn to the bottom-up processes the educative effects are those that are seen as so important in participatory democratic theory (see, for example, Holden, 1993: ch. 3, section 2). Participatory theorists regard the way in which people are developed by active engagement in political decision-making as one of the great virtues of democracy.[19] A central aspect of this is the knowledge and understanding of public affairs that is generated among the people. (There is also an important overlap here with the argument that 'democratic participation attenuates self interest' (Ward, 1998). I looked in Chapter 1 at the argument that a virtue of participatory – and in particular deliberative – democracy is its tendency to focus attention on the public good. Such attention helps to increase knowledge and understanding of what is necessary for the public good – such as policies to combat global warming.)

Bottom-up processes, it can be contended, have the greater

educational potential. And they will play a more important part later, when it is suggested that, rather than public opinion being an obstacle to action against global warming, grass-roots demands may be a crucial spur to such action; indeed, it may be that the pressure of popular opinion is needed to overcome institutional and elite inertia. However, there is not always a hard-and-fast distinction between bottom-up and top-down processes; and both can be involved in referenda and the associated campaigning.

The argument about the educative effect of democratic processes – whether top-down, bottom-up, or both – overlaps into an argument about the necessary legitimizing functions of those processes. 'The need for greater democracy arises out of the close linkage between legitimacy and effectiveness. Institutions that lack legitimacy are seldom effective over the long run' (Commission on Global Governance, 1995: 66). Here the key point is that democratic decision-making is needed precisely because, or to the extent that, people have a tendency to prioritize their short-term interests over action against global warming. Due to the extent to which they are likely to be inherently unpopular, policies to combat global warming will breed resistance and be difficult to carry through. *Because* of this, it can be argued that successful policy implementation requires the mobilization of consent and the conferment of legitimacy. It is only if, or to the extent that, popular consent is given for policies that would otherwise be unpopular, and their enforcement thereby legitimated, that they can be successfully implemented. And it is only through the increased popular understanding brought by democratic decision-making, and the operation of a democratic process, that this consent can be achieved. Despite any appearances – and dictators' beliefs – to the contrary, this may well be one of the great advantages of democracy over dictatorship. It is certainly arguable, then, that democratic decision-making, far from being an obstacle to, might in fact be a necessary condition for, a successful response to the problem of global warming.

The relevance and importance of these arguments about democracy actually being necessary to secure potentially unpopular global warming policies can be illustrated by the popular revolt against the high price of petrol that hit Europe in September 2000. In Britain, for example, oil refineries were in effect picketed and a dramatic 'fuel crisis' rapidly developed. Of most significance here was the fact that the action was effective not because of trade union muscle but because it had massive popular support. Now, an important factor in the high price of petrol was the high level of taxation to which it was subject (although another factor was the increased world price of oil, in the popular perception this should have been offset by reduced taxation). But an important reason for the tax being high was that this was part of a policy to help combat global warming; that is to say, a paramount reason for raising the level of tax on petrol had been to reduce the amount of carbon dioxide emitted by cars, by discouraging people from using them. This revolt rocked the government, and in effect it backed away from the policy because of its unpopularity.

It might seem paradoxical to claim this as an illustration of how democracy[20] is a necessary condition for, rather than an obstacle to, action to curb global warming: on the face of it precisely the reverse seems to be true. The point is, however, that the environmental case for the high petrol tax largely went by default: during the crisis the government's fundamental argument was that the tax was necessary to maintain public services. This is not to say that the environmental case would necessarily have won the day. But it is to suggest that it may have done so, if it had been put. To develop the point more strongly, it could well be that *only* an appreciation by the mass of the people of the seriousness of the problem of global warming, and the consequent need to reduce car use, could gain public support for otherwise very unpopular policies.[21] Now, according to my previous argument about the educative effect of the democratic process, the best – perhaps the only – way to bring about such an appreciation is by democratic engagement in policy-making on global warming. Equally important is the argument about the mobilization of consent and the conferment of legitimacy. Indeed, the crucial lesson of the September 2000 fuel crisis seems to be that there is little or no chance of implementing policies such as taxing petrol to discourage car use so long as they are seen simply as *the government*'s policies. That is to say, as long as such policies are seen in terms of 'them', the government, telling 'us', the people, what we must do, they are unlikely to be successfully implemented. Such success can only come if, or when, they are seen as the people's own policies. And this requires that they be decided on by the people.

These arguments against the guardianship position on securing the promotion of long-term interests in action against global warming are complemented by another. This relates to the uncertainties, mentioned earlier, concerning the benefits of such action. The essential point is this. Even though it is maintained that action against global warming *is* in people's long-term interests, the uncertainties and vagueness regarding the long-term benefits mean that there is need for judgement about just what the benefits may amount to and, correspondingly, just what short-term sacrifices are justified. And, to pick up on an argument in the last chapter, there are good democratic arguments for saying it is the people who should make such judgements. In other words, while it is a given that policies to combat global warming are in the long-term interests of the people, just what particular policies will serve those interests can only be decided by them. This argument complements those just outlined since it shows how the people's comprehension of the nature of their long-term interests, through the educative effect of the democratic process, is supplemented by the extent to which they themselves determine just what in particular is to be involved in promoting those interests.

The final pro-democratic argument to be considered here specifically brings in the idea of global democracy. This takes up another perspective on the propensity of people to focus on their short-term interests. Part of the reason for this – or at least an aggravating and parallel factor – is the way in which people's thinking is presently structured by the states system. An aspect of this system is, of course, the prioritizing of national interests, which are in important senses narrow and short-term interests. Indeed, as I shall discuss later, a crucial aspect of the problem of responding to global warming is precisely that the states system – by focusing attention on states' particular, essentially short-term, interests – tends to preclude proper recognition of the importance of general (or global) long-term interests, such as responding to the threat of global warming. The thinking of the inhabitants of states is, then, very much structured along these lines. Indeed, their very identity is normally bound up with their being citizens of states. Connected with this is the fact that the idea of 'the people' is bound up with the states system. 'The people' of a democracy are normally thought of as the inhabitants of a particular state and 'the interests of the people' are conceived in terms of narrow states' interests. On the other hand, if, or to the extent that, *global* democracy were established 'the people' would be the inhabitants of the world and their interests would be that which was in the interests of the world. The thinking of all would be structured towards global, rather than short-term narrow, interests. In other words some of the potential deficiencies of democratic decision-making as a process for furthering people's long-term interests in dealing with the problem of global warming would be overcome by more and better democracy. These matters will be taken up in the next chapter.

I have argued that the guardianship contention that people will not act to promote their long-term interests in securing action against global warming is wrong; and, indeed, that democracy is likely to be *better* than guardianship in securing such action. But I must now take up another dimension of the issue of long-term interests. My argument so far has been about the long-term interests of existing people. However, there is the further question that, after mentioning it earlier, I put to one side. This concerns the interests of future generations, and it is to this matter that I shall now turn.

The interests of future generations

Consideration of democracy, global warming and the interests of future generations raises a number of important and extremely interesting issues. They are also issues which are somewhat complex, and there is insufficient space here to consider them fully. I have discussed these matters in more depth in Holden (2000c), and here I shall restrict myself to a broad outline of salient points.

I remarked earlier that the issues regarding the worth of the long-term benefit of tackling global warming, and of whether people will act to secure it, are complicated by the extent to which that benefit will accrue to persons who do not yet exist – to future generations. My guiding questions in this chapter have been whether the mass of the people will pursue the long-term benefits of acting to curb global warming and whether, therefore, policies to combat global warming are likely to result from democratic control of the policy-making process. So far I have treated this as being a question relating to the extent to which individuals will prioritize *their own* long-term interests over their competing short-term interests. However, as I remarked earlier, it may well be that the time scale is such that the benefits of combating global warming are 'long-term' in the sense that it is primarily those who are yet to be born who will enjoy them. My central question, then, would now seem to be whether, or the extent to which, individuals will prioritize the interests of future generations over their own competing interests.[22]

Democracy and obligation to future generations

In fact matters are a little more complicated than this, and it might be better to say that a cluster of questions arises here. To begin with there is a subtle interconnection between the questions of whether existing individuals *will*, and whether they *ought to*, prioritize their own interests.[23] The latter question raises issues concerning our obligations to future generations, relating to which there is already an important body of thought and quite an extensive literature. And this body of thought contains ideas about the representation of future generations and the existence of an 'intergenerational community' that suggest the notion of an 'intergenerational democracy'. This, in turn, suggests the possibility that democracy may facilitate rather than hinder the promotion of policies on global warming that, from one perspective, can be seen as benefiting future generations rather than the present one. I will now look at some of the issues that arise here.

Let me start with the argument that democracy *will* hinder the promotion of policies designed to combat global warming. This is based on the contention that it is unlikely that people now living would be prepared to make sacrifices which would not be for their own long-term benefit but purely for the benefit of future generations. This does seem to imply (at least in the absence of a theory of 'intergenerational democracy') that democracy must necessarily ignore the interests of future generations:

> The problem is . . . that there would be almost no material compensation for people if they were to sacrifice their luxurious way of living. Moreover, those who benefit from a radical ecological policy with regard to the climate problem are future generations. But future generations are not part of the electorate. Thus one can assume that politicians risk losing their power should they try to weaken those policies that benefit the present. Producing policies that are mainly beneficial in the long run is a form of electoral suicide. (Stein, 1998: 426)

There do, indeed, seem to be good reasons for deeming it unlikely that the present generation will act against their interests purely for the sake of later generations. But, in part at least, it seems unlikely because, or to the extent that, we assume this to be a matter of self-interest pitted against charity. It is widely held that (unfortunate though this may be) people will normally only act charitably – if at all – to the extent that their self-interest is not thereby unduly harmed. But what if there is a moral *obligation* involved? Clearly, it is true that people do not always act morally. This is a general truth that applies in this particular case: for instance Al Gore (1992: 170) complains that 'instead of accepting responsibility for our choices, we simply dump huge mountains of . . . pollution on future generations'. However, Gore also seems to suggest that if not misled by faulty political processes people would accept their responsibility – and this is clearly a moral responsibility. And, generally, it is surely true that people are far more likely to act to fulfil their moral obligations than to engage in 'mere' acts of charity. And this contrast is most striking in cases where the relevant actions involve conflict with self-interest. The question that now arises, then, is whether members of the present generation are under a moral obligation to make sacrifices for the benefit of future generations (it is noteworthy that Gore maintains that they are).

Now, there is a growing literature and body of thought concerning theories of 'intergenerational' justice or equity, and of our obligations to future generations – with the latter typically being grounded in the former. It is likely that this movement of thought will increasingly percolate into popular as well as official consciousness, and help to establish a willingness to contemplate action which, while presently disadvantageous, would be beneficial for future generations. (It will further be argued below that this literature also contains or suggests some ideas that give a new dimension to the issue of democratic decision-making and the interests of future generations.) Another important consideration here is that the general notion of obligations to future generations has developed recently because those now living, for the first time, have the capacity to cause grave harm to future generations. So long as their well-being could be seen essentially as a function only of their own activities, future generations – who are, after all, temporally and therefore existentially disconnected from us – could naturally appear as morally disconnected. But the capacity to cause grave harm brings with it the moral obligation to refrain from actually causing such harm.

It is true that in the case of global warming it is (probably) not the continued existence of the human race that is at stake. Nonetheless, according to many accounts the results of unchecked global warming could be extremely harmful to most of mankind;

and, indeed, there is growing alarm about the damage likely to be caused by climate change (see the 2001 report of the Intergovernmental Panel on Climate Change, referred to earlier). But the crucial point is that with global warming, as with nuclear weapons, we have the new kind of situation where the fundamental well-being of future generations is in the hands of the present generation. Indeed, in some ways the problem of global warming has replaced that of nuclear weapons as the touchstone of humanity having a proper responsibility for its own future. And today 'the most important element in the question of intergenerational justice is the environmental issue' (de-Shalit, 1995: 7). There is an obligation to act justly in respect of those who cannot help themselves. As the 1995 Intergovernmental Panel's report on the Economic and Social Dimensions of Climate Change put it: 'Climate policy raises particular questions of equity among generations, as future generations are not able to influence directly the policies being chosen today that will affect their well-being' (Arrow *et al.*, 1996: 130).[24]

It may be that the idea that the present generation has obligations to future generations has yet to be taken up to a sufficient extent. That is to say, to an extent sufficient to counter the self-interest of the former and to bring about a willingness to make the sacrifices necessary for the latter to benefit from a mitigation of global warming. However, as indicated, it is early days as yet, and it can be expected that with the growing impact of the idea opinions will change and there will be growing popular support for action to save future generations from the harmful effects of global warming. This will reduce the extent to which democratic decision-making might inhibit such action. Moreover, the educative effects of the process of democratic decision-making itself should also be added in. As I argued in the last section, in the case of coming to appreciate their *own* long-term interests, participation in democratic decision-making will in itself increase people's grasp of the true nature of the situation. Hence, with respect to what democracy can achieve there would be, as it were, a 'double effect'. 'Participatory education' would enhance and magnify the influence of ideas of intergenerational justice, which are in any case already of growing importance, to make democratic decision-making more likely to favour future generations.

These are already important considerations, but in addition there are important arguments suggesting that democratic decision-making is actually *necessary* for effective action where sacrifices are called for. One such argument was indicated in the previous section concerning the need for the mobilization of consent, and the need to legitimize potentially unpopular policy decisions. But this argument can be supplemented by another: rather than merely denying that in practice democratic decision-making is likely to prevent the protection of the interests of future generations, it can further be argued that democracy by its nature – at least according to certain understandings of that nature – will positively foster the protection of those interests. Deliberative democracy (which is associated with participatory democracy) involves 'collective decision-making processes' in which 'what is considered in the common interest of all results from processes of collective deliberation conducted rationally and fairly among free and equal individuals' (Benhabib, 1996: 69). This 'encourages mutual recognition and respect' (Smith and Wales, 2000: 53) which necessarily involves taking proper account of others' interests:

> Among advocates of discursive[25] democracy, it is a familiar proposition that having to defend our positions publicly makes us suppress narrowly self-interested reasons for action and highlight public-spirited reasons in their place. We must do so, at least in our public explanations, if we want to give reasons to which we expect anyone besides ourselves to assent. (Goodin, 1996: 846)

The essential point here is that it can be argued that the 'others' whose interests must be addressed ought to, and therefore increasingly will, include future generations. This follows from the previous reasoning about the growing acknowledgement of our obligations to future generations. Moreover, to flesh the point out, this can be added to the thought that the 'whole thrust of modern democratic theory is to reject arbitrary delimitation of the subjects whose interests are to be politically considerable' (Goodin, 1996: 837) and conclude that to treat the interests of future generations as other than 'politically considerable' would be arbitrary. Saward, too, has an argument about the nature of democracy implying a concern for future generations (although this may be seen as overlapping into the idea that democracy itself is intergenerational, which is taken up below): it is, he says,

> inadequate to attempt to define democracy independently of offering a convincing justification for it at the same time. Now, if democracy is justified, then its justifiability cannot be said to stop with present generations. . . . [F]or the democrat a powerful concern for the aggregative and distributive nature of preventable harms with respect to future generations is unavoidable. (Saward, 1996: 87)

It is arguable, then, that with democratic decision-making – at least in its participatory and deliberative forms – presently existing people would act to protect the interests of future generations. (It is not being presumed that in considering the global warming problem the interests of future generations ought always to have priority: there are balancing judgements to be made. And it is not being maintained that presently existing people would always put the interests of those yet to live before their own, but that in reaching decisions they would properly balance the two sets of interests.) But notice that this is a line of argument to the effect that democracy will protect the interests of *others*, of those outside itself: citizens can be persuaded to take the interests of affected *non-citizens* into account. This is, indeed, very important: it is the basic line of argument in maintaining the compatibility of democracy, as conventionally understood, with securing action against global warming to protect the interests of future generations. However, there may remain doubts or misgivings about the extent to which people will decide to prioritize others' interests over their own, even where guided to do so by moral imperatives or the processes of decision-making.

At this point, though, we might question this conceptualization in terms of 'others' and their interests. That is to say we might question whether, or the extent to which, it is right to see matters in terms of a dichotomy between presently existing individuals who constitute 'the people', on the one hand, and future generations who are outside the people – 'non-citizens' – on the other. This suggests that democracy's protection for future generations might be stronger than just a matter of how far it will protect them as (albeit deserving) outsiders. It is to these issues that I shall now turn.

Democracy and the representation of future generations

The basic thought to be considered and developed is that a democracy might be conceived as somehow including future generations; and that therefore consideration of their interests would *necessarily* play a part in democratic decision-making. We have here, then, the idea of an 'intergenerational democracy'.

The central notion is that of representing future generations. The idea of representing future generations – or, at least, their interests – in today's decision-making processes is currently becoming of some importance (Agius and Busuttil, 1998; Holden, 2000c). But it is an idea that can be construed or developed in two different ways, only the second of which properly implies an intergenerational democracy. First, it is usually taken as an idea about how the decision-making processes of an 'orthodox' or 'atemporal' democracy might be supplemented. In this first interpretation the central concern is with introducing mechanisms for ensuring that today's people, in making their decisions, consider also the interests of the people of the future. But in the second construction the focus is on the notion of representing future persons as one of the elements of an intergenerational conceptualization of 'the people'.

The primary point to make about the first interpretation concerns the purpose of providing for the representation of future generations. The central purpose in fact, which is of great importance in the case of global warming policy decisions, is that without such representation future generations would be completely vulnerable to, and unable to influence,[26] any decisions taken today, the effects of which they might find extremely harmful:

> The fact that we have the increasing power to project long-range benefits does not nullify our increasing power to cast an ever-lengthening shadow of risk. And it is a shadow that increasingly falls across populations who have no say in the decisions that affect them. They have neither electoral voice, nor bargaining power nor sword to rattle. Their only avenue of representation is through our well informed concern. . . . In this context, the Maltese proposal[27] [for representation of future generations] is absolutely right. (Stone, 1998: 79)

There are other issue besides, which I shall not have space to consider here but which I take up in Holden (2000c). These concern the mode of appointment and operation of such representatives; whether it is future people, or only their interests, that can be represented; the extent to which the preferences of future persons can be known; and the extent to which remote as well as more proximate future generations can or should be represented. There is one further issue, which I shall take up here, and which leads me on to the second interpretation indicated above. This is the question of the sense, if any, in which a system of representation of future generations can be said to be *democratic*.

So let me now turn to this question. Since the advent of representative democracy there has been a temptation to see a system of representation as being *necessarily* a part of a democracy. But, of course, this is not the case – as illustrated by the fact that such systems pre-date representative democracy (Holden, 1993: 59–60). So a system of representation of future generations is not necessarily one that is democratic.

Giving an answer to the question of whether it is democratic is greatly complicated by the fact that, as we have seen, the basic idea of representing future generations can be construed in two different ways. These yield two different conceptions of what it might *mean* to say that a system of representing future generations was democratic. Under the first conception, the issue of whether a system of representing future generations is democratic seems to turn on key elements of the relationship of the representative(s) to contemporary people – elements such as the representative(s)' powers and his/their method of appointment. For example, in a democratic system the representative(s) would be appointed to a post or role within a democracy[28] by a democratic mechanism – say by popular vote or by a democratic legislature. Behind this lies the idea that a democratic system of representing future generations is one in which present-day people decide what the representative(s) shall do.[29] In a word, I might say that here the system is democratic in so far as it is grafted on to, and supplements, 'orthodox', existing, democracies. And I might call this the 'orthodox' sense of what it means to call such a system democratic.

Democratic control by future generations?

We can see, then, what, in the orthodox sense, it might mean to say that a system of representing future generations is democratic. However, it may well be that by missing a crucial point this sense is misleading. And this brings me on to the second construction of the idea of representing future generations, together with the second conception of how that idea involves a system that is democratic. The crucial point is that I am indeed talking of representatives *of* future generations – persons (or institutions) that act on behalf of *future* people. It would seem, then, that the extent to which such a system is democratic turns not on the relationship of the representatives with contemporary people but with the people of the future. It may be that while the nature of this relationship is a crucial factor in determining whether the system is democratic, it is not the only one. In a democracy representatives have to consider not just the interests of their constituents but also those of all sections of the people. There are issues yet to be addressed concerning the relevant conception of 'the people', but if present-day as well as future people are to be included in the conception, then those representing the latter must also have an appropriate relationship with the former. I return to this issue below. But if I ignore this complication for the moment I can say the idea being raised here is that a democratic system is one in which there is a democratic relationship between future people and their representatives. In an important sense this does seem to be a more logical conception than the 'orthodox idea', just indicated. A 'democratic relationship' is one where, in a significant sense, it is those who are represented who decide, at least in broad terms, what their representatives shall do. This is central to the theory of representative democracy (see note 29). But it raises two important questions.

The first of these relates to the term 'the people' and whether it can properly mean, or its meaning properly include, future people. A democracy is a system in which the people decide: they make 'the basic determining decisions' (Holden, 1993: 8). And in the theory of representative democracy the phrase 'those who are represented' is usually construed as a reference to 'the people' as normally understood. But in the (alleged) conception of democracy I am here considering this phrase is used to refer not, or not only, to present-day people but to future people instead, or as well. This might be taken to imply that this (so-called) conception of democracy is invalid. But there is an alternative implication, viz that it is acceptable to go beyond the normal, or orthodox, understanding to include as valid a meaning of 'the people' in which the reference is to future people. A further possibility is that the meaning of 'the people' should include both present-day and future people. I shall leave these issues unresolved for the moment, but the case for extension of the usual or orthodox meaning to include future as well as existing people will be considered below, when I look at the theory of 'intergenerational democracy'. And it might be noted here that the idea of extension as such – the idea that it is valid to extend the orthodox meaning of 'the people' – is crucial to the concept of global democracy that I take up in the next chapter.

A vital issue in considering the case for including future persons in the meaning of 'the people' raises my second question. This concerns the idea of 'decisions' being made by future people. The question is this: although decisions can be made *for* future people, can it also be said that decisions may be made *by* them? If only the former is the case then although there can be a system for representing future generations, it cannot be one that is democratic. And – the other side of the coin, as it were – since 'the people' are those who in a democracy make (at least the basic) decisions, then those who cannot make decisions cannot be included in 'the people'.

In responding to this question let me start from the point that, clearly, future persons cannot make decisions *now*. But does this mean that they cannot in any way be said to make decisions in a relevant sense? They will, of course, be making decisions in the future. It may be objected, though, that this is irrelevant since the very fact that such decisions are temporally separated from any being made by people today means that the two sets of decisions cannot literally be part of one and the same decision-making process. And this is important, it might be urged, because in a democratic system people must consider not only their own interests (and representatives not only their own constituents' interests) but must also engage with each other (and representatives with each other) in considering the good of, and making decisions that can be seen as being by, the whole people. But this objection may not be decisive. It assumes what has yet to be demonstrated – that 'the whole people' should include both present-day and future people. And after all, those living at any particular point in the future can themselves engage with each other in one and the same decision-making process.[30] We could focus on this process rather than either that of present-day people or the phantom idea of a process combining present-day and future people.

The idea of it being relevant that future generations *will* make decisions is, however, subject to three further objections that might be seen to have greater force. The first follows from the fact that there will be a very great number of future generations. This means there will be no 'particular point in the future' – and hence no one future decision-making process – on which one can focus. (Another aspect of the 'great numbers' argument is taken up below.) However, this objection may not apply in the case of global warming policy: it could well happen that the initial incidence of the harmful effects of global warming will be concentrated into a relatively short period of time, so that it may in fact be one particular future generation that is especially affected – in the sense of it being that generation which experiences the *change*, the deterioration, of conditions. The latest evidence suggests that it may well be the next generation that suffers this change (if indeed it is not already upon us). There is also a different kind of reason for suggesting the objection's inapplicability. This is to the effect that it is not in any case relevant to focus on a particular time and generation when considering the effects of global warming. Whether or not it happens that the change of conditions is concentrated into the period of one generation, it is *all* future generations who will suffer the harmful effects of global warming. It may therefore be assumed that the decisions of all future generations regarding their preferred global warming policy[31] would be the same. And this could be interpreted to mean that we can focus on just one future decision-making process – i.e. that of *any* one future generation.

There is, however, an alternative interpretation. Paradoxically, this brings me to consider the second *objection* to the idea of it being relevant that future generations will make decisions. According to this interpretation, if we in any case know what its results would be, then we do not actually need to conceptualize a decision-making process; and this might suggest it is irrelevant that future generations will make decisions. In fact, this is a misleading interpretation since the fact of the decision cannot be dispensed with: there is a difference between saying we know that there will be a decision to do *x* and saying we know *x* will happen anyway, without a decision to do it. And though some might see a fundamental tension between the element of unpredictability and indeterminacy central to the concept of a *decision*, and the idea of knowing in advance what will be decided, there is not really a problem here. Predictions about the outcomes of the relevant decisions can only be conceived as very probably, and not as certainly, correct. Just as when we say we know that cutting the price of some items will result in customers deciding to buy more of them, the correctness of our prediction of increased sales is very probable rather than certain. (It may be that we should talk of 'confidently predicting' rather than 'knowing' what will be decided.)

This second objection to seeing it as relevant that future generations will make decisions thus fails. But there is a third which is much more troubling and which brings me back to my starting point, the truism that future generations cannot make decisions *now*. This concerns the illogicality involved in the very idea of decision-making by future generations on matters already decided before they came into existence (see also note 31). This does seem to make a nonsense of the idea of future generations making relevant decisions. Whatever decisions they might make they cannot be ones which enter into the determinations of policy which took place *before* such decisions are made.

This surely does amount to a decisive objection to one kind of idea. This is the idea that the fact of decision-making by future generations could be relevant to formulating a conception of what kind of body is actually and literally capable of making pertinent[32] decisions on global warming policy. If such a body is to be 'the people', it can only be a body consisting of present-day people. Whether the restrictions this conclusion imposes are as decisive for possible conceptions of representative democracy as they are for possible conceptions of direct democracy is a matter I shall take up in a moment. At this point, though, it would seem that there cannot be a valid meaning of 'the people' in which there is a reference to future people.

There is an overlapping contention that reinforces this conclusion; and this is the third objection to regarding the fact of decision-making by future generations as relevant in assessing possible conceptions of the people. This concerns the 'weaker ontological status' of future generations. The contrast is made between, on the one hand, the actual existence of present-day people and their decisions, and on the other hand the potential, but as yet non-, existence of future people and their decisions. And it is held that proper consideration of this contrast shows that only the former can be included in a meaningful conception of 'the people' in a democracy.

Is it the case, then, that these arguments dispose of the idea that there can be a democratic system of representation of future generations? (It should be remembered that I am here considering only the second construction of the idea of representing future generations; but then again in an important sense this *is* the more logical construction.) At first glance it does seem that this must be so: if the argument that there cannot be a valid meaning of 'the people' in which there is a reference to future people is correct, then this does indeed seem to dispose of the idea that future people can be democratically represented. However, at this point we might look again at this argument and question how far, and in what senses, it is correct. It is true that the invalidation of the incorporation of future people into the conception of 'the people' would apply in the case of direct democracy. But my concern here is just with representative democracy, so the crucial consideration is the extent to which it applies in this case too. And, although the most straightforward assumption would seem to be that the nature of the conception of the people in representative democracy must be the same as it is in the idea of direct democracy, it is one that can be questioned. Under this straightforward assumption the difference between systems of direct and representative democracy does not involve differing conceptions of 'the people', but the presence in the latter case of an additional decision-making body – the representative assembly. Representative systems are conceived as remaining democratic to the extent that the representative assembly's decisions are ultimately determined by the people, so that it is the people who make 'the basic determining decisions on important matters of public policy' (Holden, 1993: 8). Now, the straightforward assumption is that despite a crucial difference in their content and frequency, the *form* of such decisions must be the same as in direct democracy. It is assumed, in other words, that these are decisions (albeit relatively infrequent and unspecific) that are still 'actually and literally' taken by the people – typically at elections. If this is so then here, too, 'the people' cannot be conceived as including future people.

The questioning of this assumption does not deny the existence or central importance of such decision-making in representative democracies. Rather it would ask whether this can be the only valid account of how it might be said that the basic decisions are taken by those who are represented. If other accounts are possible then this might imply that *all* of those who can be represented (i.e. including future people), and not just those who could constitute 'the people' in a direct democracy, could be said to make basic decisions. And if this were the case future people could not be excluded from valid conceptions of 'the people' on the grounds that they cannot make relevant decisions. Or, to put the point another way, since, through their present-day representatives future people could be said to engage in present-day decision-making, there is no bar on their being included in a valid conception of the people.

In fact the central notion here can be viewed in a more positive light: it can be seen as providing an account of how a system of representation can overcome a restriction of decision-making to 'the people' as conventionally understood. In this way, and contrary to the usual view, representative democracy could be seen as 'more democratic' than direct democracy – because it opens decision-making to a people that is wider than 'the people' as conventionally understood. This is essentially the point being made by Barry (1999: 221) when he says that

> Another reason for greens to endorse representative democratic institutions has to do with the green concern to give 'voice' to the interests of previously excluded others. [One of] the classes of affected interests, which do not at present have any direct democratic representation within the decision-making process, [which] have been identified by green theorists and commentators [is] . . . the interests of future generations. . . . [I]f the political exclusion of the interests of these . . . classes of non-citizens is held to constitute a defect in democratic practice,[33] then it is clear that representative institutions offer the most defensible and practical way of including them in the democratic process.

And Barry goes on to drive home the point about the 'superiority' of representative over direct democracy in this context:

> The appropriateness of representative over direct democratic forms is most obvious in the cases of the interests of future generations and those of the non-human world.[34] These groups cannot themselves express and publicly defend their interests. . . . It is therefore uncontroversial to suggest that the only sensible form that their inclusion in the democratic process can take is a representative rather than any direct form. (Barry, 1999: 221)

But let me return to the underlying contention that there can be an account of how those who are represented make the basic decisions. That is to say, an account that is other than, and supplementary to, the notion of them 'actually and literally' taking such decisions. The significant idea here is one that is in fact central to the very concept of representative government. This is the idea that the decisions of the representatives – which *are* actually and literally taken by them – are also in some sense to be deemed as the decisions of those whom they represent. In the case of democratic representative government this idea is 'given teeth', as it were, by ensuring that the decisions made by the representatives are in fact ones which those whom they represent *would* have made.[35] An important element in this, as already acknowledged, may be the decisions actually taken at elections by those being represented. But an equally important element is the anticipation by the representatives of what the future reactions of the latter might be – the 'rule of anticipated reactions' (Sartori, 1987: 152; Holden, 1993: 109). And here we have the idea that, although the only actual decisions are taken by the representatives, these are nonetheless not just 'deemed' to be the decisions of those who are represented, but are also the decisions they themselves would have taken.

This is an idea that has been applied in the context of 'orthodox' understandings of democracy and the people. And it is true that it is reliant here on the fact of future elections – actual decisions by the people – taking place; i.e. an element of the idea here is that the reactions which are anticipated will come at the next election. However, instead of referring to an inherent feature of the rule of anticipated reactions this could be seen as asserting a necessary condition for operationalizing it. Now, orthodox democratic theory would insist that this assertion is correct, but I am precisely questioning orthodox theory. And if (as a part of this questioning) this assertion is rejected then the idea could be fruitfully applied in the different context of representing future generations. Here it could show how future people, who cannot make actual relevant decisions, might nonetheless be said to engage in present-day decision-making through the decisions of their present-day representatives. Let me take, for example, the suggestion by de-Shalit (1995: 93) that some would

> claim that future generations are in a position to affect us because they can judge and criticize our behaviour, intentions, and actions, and evaluate projects we have created and developed: they have the power to decide how others will perceive what we have done.

Here it is claimed that the present actions of one set of (i.e. present-day) people is affected by the future reactions to them of another set (i.e. future generations). And if among present-day people there are those who have the status of representatives of future generations then their actions will be especially affected by virtue of that very status. Their decisions will in fact be influenced by the anticipated reactions of those whom they represent – a paradigmatic case of the rule of anticipated reactions.

There is, then, a case for saying that through present-day representation future generations can in a significant sense participate in present-day decision-making, and hence that they can be seen as part of 'the people' of a democracy. And certainly some commentators do adopt this viewpoint. Agius (1998: 11), for example, says that 'the setting up of an organ to represent future generations' gives 'posterity the possibility of participation in today's decisions'. And with a particular reference to the global warming problem, Sands (1998: 89–90) – who sees NGOs as performing the role of representing future generations – writes that '[a]nyone who participated in the Climate Change Convention negotiations (or the negotiation of other international agreements) will be aware that future generations participated in these negotiations through the presence of environmental NGOs'.

It remains true, of course, that although there is this kind of case for saying there can be democratic representation of future generations, the representatives are not, as is deemed essential in orthodox democratic theory, elected by those whom they represent. This need not, however, undermine the case. In orthodox theory such election is postulated as a necessary condition, rather than a defining feature, of a democratic system. And a corollary of the argument above concerning the rule of anticipated reactions is that such a postulate can be contested. It should also be noted that the example of ancient Athenian democracy shows that elections have not always been seen as a necessary feature of a democratic system.

Intergenerational democracy

Future people can, then, be conceived as part of 'the people' of a democracy. As suggested earlier, it could be said that this revised conception of the people provides a revised conception of democracy, which could be called 'intergenerational democracy'. And it is now time to say something about this conception.

The first point to note is that use of this conception perhaps suggests greater order than is often displayed in talk about democracy and future generations. As I have already had occasion to notice there is commonly a lack of focus on, or some ambivalence concerning, the question of whether representing future generations is a matter of supplementing atemporal democracy – by representing 'outsiders' through including representatives of future generations within its functioning – or of reconceptualizing democracy actually to include future generations. The term 'intergenerational democracy' might be used in a way that ignores or blurs these distinctions, but clearly it is the reconceptualization that is indicated by its use here.

The second point to consider is the relationship between the conception of intergenerational democracy and the idea of an intergenerational community. The former might be seen as drawing much strength from the latter. It might be agreed that the representation of an intergenerational people, ideas about which I have just been discussing, constitutes the basic structure of intergenerational democracy. However, this might be seen as a skeleton that needs to be given substance by being embodied in an intergenerational community. It might be this community that is fundamental, with intergenerational democracy being but one of the forms it can take. And just as an atemporal democracy may be said to require the prior existence of a coterminous community so an intergenerational democracy may be seen as grounded in an intergenerational community.

Now, the idea of an intergenerational community does have some force and thus might indeed provide powerful backing in this way for the idea of an intergenerational democracy. As Dobson (1998: 104) points out, the idea is not new: 'The idea of an actual transgenerational community is not new, of course, and its principles were enunciated with great elegance by Edmund Burke.' But it has been taken up again recently: 'Much more recently, Avner de-Shalit [1995] has picked up the challenge of describing an actual transgenerational community' (Dobson, 1998: 104). And Dobson usefully summarizes the description:

> de-Shalit argues that 'One of three main conditions has to be met in order for a group of people to count as a community. These conditions are interaction between people in daily life, cultural interaction, and moral similarity' (de-Shalit, 1995: 22).

> De-Shalit recognizes that the first of these conditions cannot be met between generations, but he suggests that the second two can be, based on the succeeding and overlapping generations. (Dobson, 1998: 104)

But the intergenerational community is not conceived as being confined to temporally proximate generations. There is an 'enduring cultural and/or moral similarity' (Dobson, 1998: 105) that extends across many generations. And it remains true for a considerable period of time that future people 'are born into circumstances that are recognizably similar to ours, and it is these similar circumstances that constitute the glue that holds the transgenerational community together' (Dobson, 1998: 105).

This idea, then, might provide powerful support for the conception of intergenerational democracy. Moreover, it can in any case be seen as reinforcing or undergirding the performance of the function of intergenerational democracy with which I am here concerned, viz the accomplishment of policy decisions today that requires today's people to make sacrifices for the sake of future generations. To return to the point made earlier, present-day people are more likely to endorse policy decisions of this type if they accept that they have moral obligations to future generations; and such acceptance will assist in the making of these decisions by an intergenerational people.[36] Now, the idea of an intergenerational community arguably supplies a persuasive case for the existence of such moral obligations. Thus the important communal obligations that de-Shalit (1995) sees us as having to future generations would help structure, and powerfully reinforce the legitimacy and implementation of, decisions in an intergenerational democracy regarding policies on global warming which involve material sacrifices on the part of the present generation.

There is, however, a difficulty with this idea of an intergenerational community, which Dobson sees as a barrier to providing an adequate account of intergenerational justice and moral obligations. And it can also be seen as undermining its support for a relevant concept of intergenerational democracy. This difficulty is the restricted time span of such a community. Although it is conceived as enduring over many generations, there is a limit. Over long periods of time cultural and/or moral similarities become attenuated and in the end 'fade away'. This would be irrelevant if our concern was with an intergenerational democracy of limited duration, but it is not. Since our concern is with intergenerational democracy and the prevention of the long-term – which includes the very long-term – harms of unchecked global warming, it is the idea of a democracy extending over a very long period of time that is relevant. We might consider an alternative approach to the idea of an intergenerational community via the notion of 'the community of mankind' – which would suggest an intergenerational community that is not temporally restricted. However, 'the community of mankind' initially yields a somewhat amorphous idea of an intergenerational community, one that has, indeed, rather less substance than the idea of intergenerational democracy itself.

There may be some basis for a more satisfactory account of an idea of an intergenerational community in the community of mankind notion (this is taken up below), but at this point the idea does not seem to supply any real grounding for a relevant conception of intergenerational democracy. But does it really need such a grounding? Perhaps the postulated ideas concerning the representation of future generations are sufficient in themselves. This is true to the extent that they supply an account of the basic structure, in the sense of defining features, of the conception of intergenerational democracy, just as a sufficient characterization of the atemporal idea of representative democracy is supplied by an analysis of the relevant ideas of representation. Even so, the realization of the conception may require support for such a structure, which may need to be embodied in a community. Again, there is a parallel with accounts of systems of atemporal representative democracy where there is a question about whether the basic structures of the system can be 'free-standing' or derivative of, and require the support of, an appropriate community. This is to raise central issues in political sociology concerning the extent to which political structures are the product, or a determinant, of social and cultural forms which are beyond our scope here. But it is certainly arguable that where there is a community coterminous with a democracy it is at least as true that the latter helps to maintain the former as the other way about. In fact, it may sometimes be democracy that is crucial in generating a community in the first place. Having said that, though, a coterminous community, however generated, does itself help maintain the democracy; and where such a community cannot exist the democracy is put at risk. The existence of the relevant community thus remains important for the existence of the democracy; and this could be said to be as true of intergenerational democracy.

And this brings me to another issue regarding the idea of an intergenerational community. This arises from questions about the relationship between this temporal idea and existing spatial communities. Dobson (1998: 106) sees a difficulty here: 'there is no world community in the sense in which de-Shalit would want to understand it – a community sharing a moral and cultural similarity. This means there is no transgenerational community either, just a series of transgenerational communities.' The difficulty for my argument here is that to the extent that an intergenerational democracy is to be grounded in an intergenerational community this would imply a series of geographically limited intergenerational democracies. And this would clash with my central idea of the need for a geographically *un*limited – i.e. global – democracy to facilitate curbing global warming (see the next chapter).

Now, there is, as I have just remarked, a question about the extent to which intergenerational democracy really requires this grounding. Nonetheless, as I also indicated, a global intergenerational democracy may not be possible if the existence of a relevant – and hence a global – community is prevented. And many would argue that the existence of a global community is indeed prevented: there is not simply a *lack* of a world community but, correlatively, there is the *presence* of divisions into overwhelmingly important 'partial', national,[37] communities[38] which prevent its existence. At this point, though, it should be remembered that this is precisely a difficulty on which I focus in the next chapter. And, paradoxically perhaps, a crucial argument will be the contribution atemporal global democracy could make to the breaking down, or the modification, of these divisions and the generation of a world community. It may be, then, that intergenerational global democracy can be made possible by atemporal global democracy overcoming that which is said to stand in its way – the (alleged) impossibility of a global community. In other words, the lack of a global intergenerational community, on which Dobson remarks, is a matter that is addressed in the course of, rather than something that undermines, my overall argument in favour of a global intergenerational democracy.

There is a further point to be made about the possibility of a global intergenerational community. This is that according to one viewpoint there is a sense in which it already exists. The focus in this viewpoint is on the time-honoured concept of the community of mankind. This can have interconnected temporal and spatial dimensions; and with the latter being necessarily global in extent so too is the former. For Agius (1998: 7): 'Every generation is just one link in the endless chain of generations who collectively form one community, namely the family of humankind.' Indeed, he sees the concept as currently increasing in importance and as developing specific and important connotations in the environmental field:

> Humankind as a collectivity is emerging as the new subject of rights to share the resources of the earth. . . . Another indication of the emerging notion of mankind as a collectivity may be found in the concept of common heritage introduced into international environmental law to regulate the global commons. . . . Central to the concept of the common heritage of mankind . . . is the idea of trusteeship. Certain resources of the earth are regarded as being the property of future generations as well as the present one. . . . Moreover, the idea of humankind as a collectivity which transcends the present generation is also included in UNESCO's draft Declaration on the Protection of the Human Genome. (Agius, 1998: 6)

Similar ideas had also been put forward by The Commission on Global Governance (1995: 251): 'a new need has emerged for trusteeship to be exercised over the global commons in the collective interest of humanity, including future generations'. It should be noted that the atmosphere and climate change have been specifically included in the application of these sorts of ideas. Not only do the 'global commons include the atmosphere' (The Commission on Global Governance, 1995: 251) but 'climate change [was classified as] a common concern of mankind[39] in the resolution adopted by the General Assembly regarding the Protection of Global Climate for Future and Present Generations of Mankind' (Borg, 1998: 137).

The idea of an intergenerational community of mankind is, then, of growing significance and is of particular importance in the context of the environment, especially in relation to global warming. Does this mean that there is no 'difficulty' of the sort indicated by Dobson, and that a global intergenerational community already exists? The short answer must be no. Clearly, it is the *division* of the world into different national communities that, as yet, remains basic. In the face of this the idea of a community of mankind appears inoperative. And, as already remarked, even if, or to the extent that, it is operative its content seems too amorphous to have relevant practical effect. The full answer, however, is a little more complex. The division of the world is subject to some modification by the idea of a community of mankind – whether or not with a focus that includes an intergenerational element. That there is such an idea, with a notable pedigree (see, for example, Schiffler, 1954), supplying an alternative vision of the world, is of significance in itself. But it is also one which is actually having some effect on the workings of the world, as is demonstrated by the very manifestations of the idea in international environmental law I have been considering. Further than this, it is an idea not without importance in the theoretical ferment associated with the current period of global change – a ferment which becomes a factor in the process of change itself (see the next chapter). A global community has yet to be generated, but the idea of the community of mankind has some importance in the processes (which include the development of atemporal global democracy) that might lead towards this. Moreover, there is some reciprocity here and the kind of global community that may be in the process of development can in turn serve to give greater substance to the notion of a community of mankind. And it is arguable that this is the kind of substance that has not only a temporal dimension, but one which would 'carry over' to remote generations. This would make the community of mankind idea in the end supportive of an idea of intergenerational democracy with a relevant – i.e. very long – time span. Such an argument cannot be developed here, but it would centre on the interaction and mutual support between the vision of humankind as a meaningful entity with infinite (or, at least, indefinite) temporal existence, and the existence of ongoing structures and processes which – by to a significant extent integrating the myriad, far-flung and disparate groups and individuals in the world – give some concrete substance to the idea of 'humankind'. And an implication of such an argument would be that such structures and processes could have a 'non-fading' substance and continuity that would not 'fade away' in the same way as de-Shalit's constituents of an intergenerational cultural community.

Conclusions

Let me now try to draw some threads together and focus on the general nature of the analysis and its concern with the representation of future generations and the idea of intergenerational democracy.

The interest in these themes grew out of the basic concern with the question of the extent to which democracy might inhibit or facilitate action to curb global warming. In the previous chapter I discussed the capacity of the mass of the people to make or endorse the policy decisions necessary to combat global warming. I rejected the argument that ordinary people cannot know what serves their interests in complex matters like the global warming problem, and only experts can. I then moved on, in the first part of this chapter, to consider the argument that ordinary people will nevertheless not support the policies necessary to curb global warming because they are incapable of discerning and/or pursuing their long-term, as opposed to their short-term, interests. This in turn was rejected, but I then had to confront another dimension – or extension – of the argument. This was to the effect that the relevant 'long-term interests' are those of future generations, and hence the issue became one of the extent to which ordinary people would promote the interests of future generations at the expense of their own.

In confronting this issue, however, it was posited that the question of whether the interests of future generations would be promoted by democracy need not simply be one of the extent to which those interests would be promoted by the unstructured decisions of present-day people. Instead it was suggested that democracy could incorporate the representation of future generations so that the advocacy of their interests could be embodied in the very structure of decision-making in a democracy. I then raised the question of whether this might mean the supplementation of democracy, by the grafting of 'alien' elements on to orthodox or 'atemporal' democracy, or its reconceptualization as 'transgenerational' or 'intergenerational democracy'. And I suggested the latter might provide the more satisfactory answer.

Use of the concept of intergenerational democracy means, then, that it can be more convincingly argued that democracy would not obstruct action to protect the interests of future generations. Indeed – in the light of the analogous argument regarding atemporal democracy – it can be argued it would positively facilitate such action. And this, of course, greatly strengthens my overall argument about democracy and global warming, both in its 'negative' form, to the effect that action would not be inhibited by democracy, and its 'positive' form, to the effect that it would be actively assisted.

Central to the argument about the concept of intergenerational democracy's contribution to action against global warming is the way it provides for the interests of future generations to be integrated into the very functioning of democracy. And this is, indeed, the idea which has been central to my concern with intergenerational democracy. But before I move on there are some points which arise here that require comment.

The initial point is that considering the interests of future generations does not mean ignoring the interests of the present generation. Rather, it means balancing the interests of the present and future generations.[40] Now, presumably, as in an atemporal democracy, it should be said that such balancing would be done democratically. (It is noteworthy that de-Shalit (1995: 58–9) says that 'the precise decision as to how much of what and when to allocate [to different generations] must, I believe be taken . . . democratically'.[41]) But this raises at least two issues or difficulties. True, one of these has already been discussed – the difficulty with the notion of a 'decision' in this context. The other, however, has not. This concerns relative numbers and the implications of the democratic principle of 'majority rule'. (Although this principle does itself have a problematic place in the theory of democracy – see, for example, Holden (1993: 43–7) – it remains the case that the idea of democracy is normally seen as in some way giving greater weight to greater numbers.) Clearly there will probably be more individuals in future generation*s* than in the present *one*. And the more generations are included, the greater the imbalance. Odd and repugnant conclusions might seem to follow from this – using orthodox ideas regarding the majority principle – about the domination, rather than the balancing, of the interests of future generations. Now, apart from issues regarding the status of the majority principle itself, there may be some pretty intractable issues here. However, it does seem reasonable to maintain that the fundamental divide is that between the present and all future generations (see note 40); and it can be argued that a corollary of this is that those on each side of the divide should have equal weight. This implies, of course, that future generations are not to be given greater weight because of their greater numbers, but this can be seen as a valid way of balancing out their 'weaker ontological status'. The intuitive thought here is that the present non-existence of those yet to be born means that they have less importance. (This is not to say they have no importance: indeed my whole discussion of taking account of the interests of future generations is premised on the assumption of their importance. Rather it is to suggest a reasonable way of dealing with the undoubted significance of the fact of the real existence of members of the present generation as actual individuals.)

In conclusion it can be said that the concept of intergenerational democracy greatly strengthens the argument that democracy has a beneficial role to play in action against global warming. The strengthened argument draws on the contention that an atemporal understanding of democracy can logically be completed, as it were, by an intergenerational understanding, and that this could naturally, and very beneficially, be grafted on to the traditional atemporal concept. But there is more to it than this. Or rather, the 'grafting on' is not just a matter of an extension of the idea of atemporal democracy, important though this is, since it also, and crucially, involves adding actual structures for *securing* the representation of future generations.

These potentialities of democracy, then, make it uniquely suited to bringing about the kinds of decisions, focused on very long-term interests, that the combating of global warming requires. All that is needed now is to 'complete' the spatial understanding of democracy, since the kinds of decision required are not only long-term, but also global. I shall turn to this in the next chapter.

Before I do this, however, there is one final point concerning intergenerational democracy to be considered: it might be objected that the concept as such has no necessary relevance for democracy itself. The key argument is that mere employment of the concept would not in itself entail the *existence* of intergenerational democracy, any more than, say, the use of the concept of 'unicorn' entails the existence of unicorns. Now, this argument may be overstated. Concepts relating to social reality can themselves have an effect on that reality in a way that concepts relating to non-social reality do not. But, although there are some profoundly important issues here, at the end of the day since the effect of theorists', rather than participants', concepts may be minimal I shall treat this argument as valid.[42] This does not mean, however, that the objection is upheld. Again, there are some complex and important issues, but the essential point is that even where concepts do not themselves directly affect social reality clearly they may nonetheless do so indirectly in the sense that actors – politicians, activists and so on – may seek to change that reality to accord with a concept or set of concepts.[43] In this way reality could come to be shaped by the concept of intergenerational democracy. What this could mean in practice is that the actual behaviour of present-day people would be shaped by their perception of being part of an intergenerational democracy. Part of this would consist in relatively unreflective behaviour structured directly by institutions for representing future generations. But to an important extent individuals' behaviour could also be shaped by their cognizance of being members of an intergenerational democracy and of having an appropriate role to play therein (this would extend beyond, but include, reflection on the nature and purpose of the institutions for representing future generations). In short, the logic of the concept of intergenerational democracy – and in particular its incorporation of the interests of future generations into decision-making – could translate into actual patterns of behaviour. And that, of course, is why the concept is important.

Notes

1. 'Benefit' needs to be understood in a rather negative sense in the context of action against global warming. Since it is only abatement of global warming, and not its prevention, that is now possible, the benefits of action will consist in suffering consequences of global warming less adverse than they would otherwise have been.
2. Portney and Weynant are in fact concerned with a cost–benefit analysis of global warming abatement in order to deal with the issue of 'discounting' (in the economists' sense). That is to say, their concern is with how to give an economic measure today of the future economic benefits of abatement that will enable a calculation of the economically appropriate rate of investment to provide for that abatement. However, this does raise some of the same basic questions of balancing present and future costs and benefits that we are concerned with (though with the qualification that one of our issues is the extent to which these may not – or may not only – be *economic* costs and benefits).
3. Schelling is in fact here concerned with some implications of the argument that the future beneficiaries will be in the now developing countries while those who would now forgo consumption are in the already developed countries. I shall for the moment sidestep the cluster of issues this raises (I take them up in Chapter 4) to deal here with the matter of the relationship between present and future costs and benefits for the same group of (initially) the same individuals.
4. In a note (38) Kavka and Warren add: 'It is entirely rational to discount

expected future benefits in virtue of the uncertainty of their receipt, or of one's later need for them, or of one's still being in a position to enjoy them. Discounting beyond this, absent further considerations, would constitute having an irrational "pure time preference".' (Some of the issues raised by the first case in respect of action against global warming I take up elsewhere, but essentially it is held here that the future benefits of action are sufficiently certain to rule out rational discounting on these grounds. The second and third cases similarly fail to yield rational discounting here.) Aspects of the issue of the rationality, or otherwise, of pure time preference in the case of combating global warming are taken up below.

5. To avoid circumlocution the point is stated in this 'back-to-front' manner. The phrase 'the benefit brought by not acting to curb global warming' is perhaps more naturally understood as meaning 'the benefit brought by engaging in activities that would be circumscribed by action to curb global warming'.
6. There are some complications here since the people now living are of different ages. This of course means that some older individuals may not experience anything of a delayed benefit, though it may begin to occur in time for a part, at least, of it to be enjoyed by those who are younger: and the youngest of the latter would experience more of the benefit. For the sake of clarity, however, I shall not here attempt to cover all these permutations. The essential points can be indicated by talking simply of 'existing individuals' or 'individuals now living': such phrases could be taken as referring to a generation, or the cohort of people of some median age range. (There are, though, possible conceptions of 'the people' which incorporate future generations, for which it is of some importance that individuals who co-exist are at various different points in their lifespan.)
7. This argument will arise where it is maintained that correctness cannot be determined, or is not an issue, either in particular cases or as a general epistemological position.
8. What Richard Wollheim (1962) called the paradox of democracy is that a democrat can have contradictory beliefs about what ought to be done: he may think *x* ought to be enacted because it has been democratically decided upon, but at the same time he might think y ought to be enacted because it is the better policy. For further discussion and references see, for example, Harrison (1993: 223–7, 243). One of the reasons for favouring the democratic decisions even if they are 'wrong' is, of course, that the case against guardianship does not rely solely on arguments about the competence of the people. As pointed out in the last chapter, even if it is only the guardians who have the competence they can be corrupted by power. In any case, here I shall be arguing that people *are* capable of pursuing their true interests; and if the argument is successful the paradox does not arise.
9. I am indebted to Richard Bellamy for this suggestion.

10. The latest and most authoritative warnings to this effect are to be found in the 2001 report of the United Nations Intergovernmental Panel on Climate Change (IPCC, 2001).
11. Evidence of this sort can clearly be used to bolster the guardianship argument. As against this, however, there are five points to make. First, there are still many among the elites who do not treat global warming as a proximate threat. Second, even if such a view has not as yet gained much ground among the people as a whole, it may yet do so. Indeed, as already indicated I shall argue that engagement in a democratic decision-making process about global warming would itself have this effect (see below in the main text). Third, the question of whether or not the harmful effects of global warming are proximate, and who perceives them to be so, is peripheral to my main argument, which is primarily concerned with perceptions in a scenario where there *is* a significant time lapse. This 'worst case' scenario provides the most rigorous test of the 'pro-democracy argument' that guardianship is not needed for combating global warming. Fourth, even if some harmful effects are already occurring, or will do so very soon, it is probable that the occurrence of the most harmful effects is still some way off; and since it is upon these that perceptions of the dangers of global warming tend to centre, the critical scenario remains the same. Fifth, policy-making elites are overwhelmingly national elites, concerned with the national interest. Combating global warming requires global action, and arguably this requires global democracy. This is the argument which will be developed in Chapter 4.
12. The usual perception is that the necessary reductions could only be brought about by measures of governance. However, such measures need not necessarily take the form of 'direct regulation', but can consist in setting up an appropriate economic framework. Indeed, there are growing literatures on (a) reformulating concepts of economics to obtain measures of economic value that take proper account of environmental goods; and (b) using taxation policies and economic incentives to encourage environmentally friendly activity (for example, 'Pearce economics' – see, for instance, Pearce (1992) and Pearce and Barbier (2000).) This flows over into what is now a growing interest in the possibilities of economic activity *rather than* governance. Consider, for instance, the extent to which big business (for example the insurance industry and elements within the oil industry) is beginning to see tackling environmental problems as being in its interests. One important line of argument is to the effect that with an appropriate redirecting of economic activity there need not be any reduction in material standards of living. (And this would obviate significant aspects of the clash between short- and long-term interests. See also the second point in the main text above.) Some would argue that it is in this area that the most realistic prospects for combating global warming are to be found. The present writer thinks that this is over-simplistic

(while acknowledging there are important issues here) and that issues of (global) governance and the regulation of the pursuit of short-term interests remain central. In any case, as explained in the main text, there is a more severe test of the pro-democratic case regarding the global warming problem if it is accepted that there is a genuine clash between short- and long-term interests.

13. It is possible to 'stand the argument on its head' here and say that a successful democratic response to global warming is unlikely or impossible precisely because it requires the dissolution of the capitalist system. Or, to put the point another way, because the dominance of the idea of a material standard of living is a product of the capitalist system it is unrealistic to suppose people will come to conceive living standards, and therefore their short-term interests, in a non-materialist way. The dominant idea may be false, but it is – and will remain – dominant because people are the victims of 'false consciousness'. Involved here, of course, are underlying assumptions about the entrenchment of the capitalist system, which now must mean – and this is especially relevant for global democracy and global warming – the global capitalist system. Essentially, then, what are raised here are the large and basic questions concerning democracy's capacity to prevail in the face of capitalism, which are beyond the scope of this book. In any case my argument about global warming and democracy will involve affirming the ability of people to look beyond their short-term interests, and so will not require those interests to be non-materialistic. Capitalism, in other words, will not be seen as a bar to popular support of global warming abatement policies – and will not therefore be an issue (unless arguing for the ability of people to transcend short-term material interests is itself taken as an argument for democracy's capacity to transcend capitalism). From another angle one might say my argument avoids utopianism by considering prospects for action against global warming that do not involve the abolition of global capitalism.
14. Just to take one example of this, the report in June 2000 of the Royal Commission on Environmental Pollution warned of the need 'for deep cuts in the use of fossil fuels and radical changes in lifestyles' in order 'to avert environmental catastrophe' (*Independent*, 17 June 2000).
15. Arguably, an important reason why people tend to focus on their short-term interests is the 'distorting effects' of the states system and the way in which this clouds perception of long-term global interests. This point is taken up later.
16. In the interests of clarity of expression phrases such as 'prioritizing long-term over short-term interests' are normally used to convey the meaning more precisely specified here.
17. There is here an echo of the guardianship position. And, indeed, it is arguable that conventional democratic theory, with its stress on the role of representatives who are wiser than their electors, constitutes a desirable

and sophisticated 'middle way' between theories of guardianship and radical democracy: see, for example, the passage in Pitkin (1967) referred to in the main text above. The difference, of course, between 'conventional' democratic theory and the guardianship position is that in the former in the end the mass of the people do make the final determining decisions concerning, and do have to be convinced by, the policies advocated by the leaders.

18. Differing perceptions here parallel, and may be linked in with, differing accounts of the relationship between 'elitist' and 'pluralist democratic theory' (in its former, narrower, sense – see Holden (1993): note 40: 120–1). Pluralist is often regarded as akin to elitist democratic theory, and some notable exponents – for example Robert Dahl – have been characterized as 'pluralist-elitist' democratic theorists. When regarded in this way, the role of groups in pluralist democratic theory is seen as akin to the role of elites. On the other hand, when groups are seen as vehicles of popular participation it is possible to see pluralist democratic theory as a form of participatory democratic theory: see, for example, Holden (1993: 130) and Kelso (1978).
19. Participatory democratic theory is critical of much of mainstream liberal democratic theory – and especially of elitist democratic theory. A central feature of this is an insistence that only a political system in which there is active and positive participation in political decision-making by the people *can* be a democracy.
20. The petrol price protest could be interpreted as an exemplification of democracy in either or both of two ways. First, since the evidence suggests it was effective because of widespread popular support, it could in itself be seen as a manifestation of the will of the people. It should be noted though that the government's line was that the action was *anti*-democratic since the democratically elected government was being intimidated by (what they saw as) a minority of protesters. But this brings me to the second interpretation. The protest provided evidence of likely electoral behaviour; and to the extent that future government policy on petrol taxation was determined (or heavily influenced) by anticipation of electoral consequences, then the will of the people was being manifested through the operation of the electoral system.
21. It may well be that it is only a more sophisticated and multi-faceted policy that could gain public support. Such a policy would probably include some provision for viable alternatives to car use and a more nuanced tax regime – perhaps with some discrimination among categories of car users (or, indeed, a regime of tax on the type of journey rather than on petrol). But this again shows the need for democratic decision-making, since this would reveal just what policies would gain support.
22. Strictly speaking the question has more facets than this. Thus I might say that, while there is a clash between existing individuals' short-term

interests and the interests of future generations, there is an overlap when it comes to existing individuals' long-term interests. To some extent this may be true. However, there are further complications regarding the uncertainty of the timing of the incidence of the benefits of global warming and the differing prospects of the different age groups of individuals now living (see note 6, above). In any case there are those who see the timing of the effects of curbing global warming to be such as to clearly benefit future generations *rather than* those now living. And it is prudent to consider a 'worst case scenario' as far as my thesis about democracy's capacity to meet the problem of global warming is concerned: it is when it is postulated that action to combat global warming involves existing people making sacrifices which will benefit not themselves but only those who will exist in the future, that the idea that those people will support such action is most severely tested. For this reason, and for the sake of simplification in the interests of clarity, in this section it will be assumed that the question is one of a potential clash of interests between existing and future generations.

23. As before, it is given by the premise of my argument that that which is necessary for the curbing of global warming is that which ought to be done. Hence, it is taken as a given here that existing individuals ought not to prioritize their own interests over those of people yet to be born, where this inhibits action against global warming (I am, of course, here staying with the assumption of this section that the benefits of such action will only, or disproportionately, accrue to future generations). And, again as before, this rules out the opposite view – that existing individuals ought to prioritize their own interests – being validated by a democratic judgement approving it. However, this somewhat stark way of putting it does conceal some complications. Not the least of these is that there is still a need for some judgement about the balancing of conflicting generational interests within the general given that future generations' interests in the curbing of global warming should be prioritized. Some of the implications of this point will be taken up later.
24. It can also be argued that instead of – or besides – relying on the just behaviour of the present generation, future generations should somehow be represented in today's decision-making processes. This point will be taken up below.
25. 'Deliberative democracy has also been termed discursive and communicative democracy' (Smith and Wales: 62).
26. The question of whether, or in what sense, (as yet) non-existent people can be said to exert influence is a matter which is taken further in Holden (2000c), and which comes up in my discussion of the second issue in the text below.
27. 'In 1992, in preparation for the 1992 Rio Earth Summit . . . the delegation from Malta submitted to the preparatory committee that the

world community go beyond . . . vague declarations of responsibilities towards future generations . . . and actually institute an official guardian to represent posterity's interests' (Stone, 1998: 65).

28. Here 'within a democracy' should be taken to include – or, indeed, to specially refer to – 'global' democracy. Discussions of the idea of the representation of future generations have, with good reason, tended to focus on global – or at least international – structures and institutions, e.g. Agius and Busutill (1998). And the central thesis of the next chapter is that success in combating global warming requires *global* democracy.
29. As shall be shown in the main text below there are some significant points to be raised about the sense in which the people make decisions. But there is, of course, also the issue of just what decisions they make, in whatever sense it is that they make them. This concerns the degree to which in a representative, as distinct from a direct, democracy the people make the policy decisions. But it is central to the theory of representative democracy that a representative system can only *be* a democracy if it is one in which the people make at least 'the basic determining decisions on important matters of public policy' (Holden, 1993: 8; see also 54–5 and 58–62).
30. Such a formulation might be taken to imply something that many might see as absurd, viz that (at the relevant point in time) *everyone* – i.e. everyone in the world – could engage in the same decision-making process. However, (a) the essential point could be reformulated to avoid this implication; (b) in any case it should again be remembered that the idea of global democracy is central to the next chapter – an idea which involves the notion of everyone in the world being engaged in some sense in the same decision-making process. Two other points should also be noted. First, there can be a flexible interpretation of 'those living at any particular point in the future' engaging in 'one and the same decision-making process' to include a process of decision-making taking place over the period of, and involving the members of, one generation. Second, the term 'everyone' should be taken to mean 'all adults': see, for example, Holden (1993: 13–14 and 49, n.7).
31. A more precise, but more clumsy, formulation might refer to future generations' decisions regarding the policy they would prefer the present generation to adopt. In fact, even this way of putting it involves an oversimplification since the reference should really be to what future generations would decide if they were able to take part in decisions made before they came into existence. The implications of the illogicality exposed here are taken up in the main text below.
32. Future generations will of course be able to make their own kinds of decisions on global warming policy, but the ambit of such decisions will be determined and restricted by the policy decisions already taken – or not taken – by present-day people. If there is to be any hope of curbing global warming, decisions have to be taken now, and in the near future. And

what future generations cannot do is take part in these crucial present-day decisions.

33. Following the terminology I have been using it might be better to talk of 'what would be conventionally regarded as non-citizens' and 'a defect in what is conventionally understood as democratic practice'. There is, in fact, some ambivalence concerning what it is that Barry is advocating. However, at least part of his argument parallels the one developed here to the effect that there could be a reinterpretation of the conception of 'the people' (though he does not *quite* follow this through): 'Although democracy is necessarily *by* the people and *of* the people, it does not necessarily have to be *for* the people, where the "people" is understood as a human community presently living within a nation-state' (Barry, 1999: 221). See also the next note.
34. As I have remarked before, Barry considers the possibility of representing the interests of '(parts of) the non-human world', but I do not.
35. In traditional democratic theory there are differing accounts of how this takes place, ranging from those of 'radical democratic theory' where representatives simply reflect the views of their constituents, to those of 'conventional democratic theory' where representatives use their own discretion but their decisions must be ones which their constituents would ultimately endorse (see Holden, 1993: 68–9).
36. In an intergenerational democracy the idea is that decisions taken today are not simply decisions taken by members of the present generation but are at the same time, in an important sense, decisions in which future generations partake since they are part of the intergenerational people. Clearly, though, present-day people will to some extent remain reluctant to accept policies requiring sacrifices by them for the sake of future people. And anything that contributes to gaining such acceptance will be helpful. This could be interpreted either as helping present-day people to support the relevant policy decisions by an intergenerational people, or as helping them to accept that they are part of an intergenerational people.
37. The point remains essentially the same if it is argued that national have now been superseded by transnational cultural communities or 'civilizations': see Huntington (1993, 1996).
38. The time span of a relevant intergenerational democracy may mean, as already suggested, that it could not be constituted by an intergenerational community. Nonetheless, its existence might be prevented by the positive prevention of the existence of an intergenerational community of the relevant geographical scale, rather as an atemporal democracy, although not constituted by a coterminous community, could be subverted by the factors making such a community impossible.
39. 'The concept of common concern of humankind, is closely related to . . . the concept of the common heritage of mankind. . . . [But] a number of states objected to the idea of classifying climate as part of the common

heritage of mankind when Malta requested its inclusion in the provisional agenda of the 43rd session of the General Assembly. Malta's intentions, however, were not to speak in terms of appropriation of any part of the climate system but in terms of protecting it. The term common heritage was therefore replaced by the words common concern' (Borg, 1998: 136–7).

40. It might be argued that the balancing required is between the interests of different – including different future – generations. However, I shall reject this for two reasons. First, in the case of global warming, to the extent that there are different interests involved – to the extent that global warming poses a significant threat only to future generations – the issue is such that the present generation has interests (in not making material sacrifices) that conflict with the interests of *all* future generations (in not suffering the effects of unchecked global warming). Second, there is a crucial difference – an 'ontological' difference (see the main text below) – that renders the distinction between the present and (all) future generations in important ways much more significant that between different future generations.
41. It may be that de-Shalit means to refer to an 'atemporal' democratic decision (i.e. by present-day people); he does not make it clear. However, his context seems to call for an 'intergenerational' democratic decision.
42. And the effect of theorists' concepts relating to collectivities – such as democracy – may be even less than the effect of those relating to individual behaviour (essentially because the latter can – in relevant circumstances – enter directly into participants' behaviour). Nonetheless, it may be that too much is conceded to the objection here: it is true even of concepts relating to collectivities that their deployment by influential theorists and commentators can sometimes affect the reality being theorized (consider, for example, the influence of theorists of the nature and role of the free market). This point is different from the one made directly below in the main text, since it concerns the effect that concepts themselves can have. However, in practice there may not always be much to distinguish the 'direct' effect of a concept itself and its 'indirect' effect through being taken up by those seeking to change reality.
43. Actors' efforts can themselves be affected – whether assisted or frustrated – by other developments. Global democracy (see the next chapter) is a good case in point: efforts to create it may be seen as frustrated by processes of globalization but arguably these processes assist such efforts. Also, globalization processes might be seen as provoking thought about, and efforts to move towards, global democracy in the first place. See Holden (2000b).

CHAPTER 4

Global Democracy and Global Warming

Introduction

My concern in this book is with the implications of democracy for action against global warming. So far I have taken the central question to be the likelihood or otherwise of ordinary people making, or supporting, decisions in favour of effective global warming policies. Of course this *is* a central question, and, because it is, it has been beneficial to focus on it separately. But this has meant sidelining the other key issue, regarding the *kind* of democracy needed if ordinary people are indeed to make effective global warming policy decisions. The question here concerns the geographical extent of democracy. Although in the last chapter I discussed its 'temporal extent', I have not as yet taken up (though I have referred to) this issue of its 'spatial extent'.

The nub of the issue is that a democracy is normally conceived as being coterminous with a state, whereas global action is required to combat global warming. In other words, given the traditional and usual use of the term 'democracy' to mean a form of state – a form of political decision-making within a state – the paradoxical implication is that our discussion so far has been concerned with a political form that is incapable of providing action against global warming. What is now required, then, is an assessment of whether it is possible to have a *global* democracy. If it is, then the argument of the last two chapters can be 'rescued': the allegation that democracy is a political form incapable of providing action against global warming would be refuted and the applicability of those arguments would be upheld. In this chapter I shall contend that global democracy is a possibility. And in fact such a contention involves adding to and sharpening – rather than simply upholding the applicability of – previous arguments.

I shall begin by looking at the factors that have led to the emergence of the idea of global democracy. I shall then consider further the nature, and the possible actualization, of this idea, and how global democracy could help bring action to curb global warming (and how it might itself be bolstered by the need for such action).

The developing interest in global democracy

One of the factors contributing to the emergence of the idea of global democracy is the current prestige of democracy itself. It is now commonplace to maintain that with the end of the cold war and the downfall of communism, there has been a 'triumph' of democracy. Despite some criticism of this idea it is now very widely accepted that democracy is the best form of government. And even if it is not accepted that 'the principles of democratic government [are] triumphing' (Hadenius, 1997: 1), it can at least be agreed that '[n]ever before has the idea of democratic government been more popular' (Archibugi, Held and Köhler, 1998a: 2). Indeed, it can be said that '[a]mong the twentieth century's most important legacies to the new millennium [is] . . . the assertion of democracy as the legitimate form of government' (Archibugi, Held and Köhler, 1998a: 1).

Now, the reference here is of course primarily to the traditional idea of democracy within a state. However, there is a growing tendency for the international pre-eminence of intra-state democracy to flow over, as it were, into ideas about, and affecting, the international order. 'It is an international lawyer who notes the new stress on democracy, as . . . "a dominant value, in national and international affairs"' (Smith, 2000: 206, quoting Crawford, 1994). But it goes further than this, so that the 'triumph' of democratic notions 'may well prove to be . . . the fulcrum on which the future development of global society will turn' (Franck, 1992: 49). International norms and practices are, then, being affected by the democratic idea. In other words, the primacy of the democratic idea *within* states is beginning to extend to the order of relationships *between* states: the pre-eminence of democracy within the components of the international order is beginning to affect the nature of the order itself.

Although such an account is by no means universally accepted, it is becoming increasingly difficult to deny that some such processes are under way. However, there are different ways of viewing these developments. On one view what is, or may be, happening is some sort of democratization of, broadly speaking, the existing international system. There is, though, a conceptual difficulty here. There remains, on this view, a fundamental conceptual disjunction between states themselves and the system within which they operate. Because of this, and because democracy has hitherto been understood as a *form of state*, it is difficult to know what 'democratization' of the *international system* could mean. Be that as it may, it could be that this whole view will in any case become outdated: it may well be that the developments in question, in conjunction with others (to be discussed below), are generating changes which are more complex and extensive than simply changes to the existing form of international system. It may be, then, that there are developing changes, both in the understanding of democracy and in the nature of the international realm, that prefigure a new form of 'global' – rather than 'international'[1] – order, that is itself democratic and which in some sense subsumes democratic states. Mulgan (1994: 188) talks of a 'long-term trend . . . towards the sovereignty of an embryonic world public'. But whether or not it is conceived quite in this way there is no doubt that there is now a growing interest in the idea, and possible development, of 'global' or 'cosmopolitan' democracy (see, for example, Archibugi and Held, 1995; Held, 1995; McGrew, 1997b; Archibugi, Held and Köhler, 1998b; for an overview see Holden, 2000a).

The changes in the understanding of democracy and in the nature of the international realm overlap and interconnect to a significant extent; it is, however, convenient initially to approach them separately. I shall look first at the former.

Changing understandings of democracy

The central point to be made here is that the core idea of democracy, rule by the people, is beginning to be rethought. An essential reason for this is in fact the need for democratic thinking to respond to some of the changes in the international order I shall look at further below. These can be seen as tending to render obsolete received notions of 'rule' and 'the people'. This is because the traditional concept of democracy centres on the state, while the changes mean states are becoming less pivotal. In the received notions the state gives content to the idea of rule by the people. It does this by postulating the state as the only unit through which there can be rule and specifying 'the people' as 'those who are citizens of the state', thereby implying that 'rule by the people' must mean 'control of the state by its citizens'. As states become less pivotal they lose power, including their capacity to be instruments of rule. Correlatively, there is a tendency, perhaps, for the activity of 'ruling' – or at least of 'governance'[2] – to move beyond the confines of the state.[3] At the same time – partly because of states' diminishing importance, and partly for other reasons – citizenship of a state is ceasing to be the sole means of specifying 'the people'. It is, then, increasingly being thought that 'rule by the people', to the extent that it exists at all, must consist in something other than – or at least must include more than – the control of states by their citizens.

Let me now consider each of these developments in a little more detail. I shall begin with changing conceptions of 'the people'.

The idea of 'the people'

I am interested in the development of what can be called 'transnational conceptions of the people'. I shall be specifically concerned with but one of these, the idea of a global people. However, since the development and credibility of this and other transnational conceptions are linked together – linked in a relationship of reciprocal causation with the undermining of the traditional or orthodox state-centred conception of the people – I shall have some concern with all transnational conceptions. But let me look first at the undermining of the traditional conception.

It could be said that there has always been a hiatus in democratic theory when it comes to specifying 'the people'. According to the traditional view, I have just suggested, the specification is supplied by the idea of the state. However, the rationale for this has been assumed rather than made specific (and of course it is precisely the existence or validity of such a rationale that is now at issue here). Until now the matter of the specification of 'the people' has usually been regarded as being beyond the province of democratic theory. Indeed, it poses a problem which can be seen as 'insoluble within the framework of democratic theory' (Whelan, 1983: 16).[4] And the 'solution' has been supplied, outside this framework, by presuppositions about the world derived from nationalism and the prevailing states system. 'The "sovereign people" came to be identified with the nation' (Heater, 1990: 56) and since, 'until recently, at least', liberal democratic theory has 'accepted as a given in a world divided into nation states' (Ware, 1992: 132), it has been 'a given' that the people is coterminous with the state. But in the newly changing world it is increasingly difficult to accept that this is 'a given'. It is becoming increasingly unrealistic to presume either that there is a coincidence between existing nations and states or that the division into states is necessarily a fundamental feature of the world. There are important interrelationships, but broadly we can say there are two sorts of reasons for this. On the one hand there are the changes in the nature of the international realm, which I shall look at below, and on the other there are changing perceptions of, and criteria for specifying, 'the people', which I shall consider now.

There are at least two important developments involved in these changing perceptions and criteria. The first is a resurgence of disruptive ethnic nationalism – disruptive, that is, of established states. Part of the reason may be that as the state, due to globalization (see below), becomes less important the nation becomes more so (Delanty, 2000: 95). In any event, in Quebec, the Balkans, the former Soviet Union and elsewhere, nationalisms are threatening, or have caused, the undermining of established states. In other instances (Northern Ireland for example) nationalism has been the cause of bitter dispute about the territories of states. And it may be – partly causing this and partly resulting from it – that the concept of the nation state is dissolving: 'there has been a decoupling of nation from state' (Delanty, 2000: 98).[5] And certainly the idea that there is a salient division of the world into nation states is being called into question. This disrupts the very tight connection between 'the people' and 'the state', since it dissolves the key basis of that connection, the assumed automatic coincidence of states and nations. This was the basis because in fact the identification of the people with the *nation* was usually primary; and the identification with the state flowed from this via the notion of the nation state.

The state specification of 'the people' is, then, being undermined by ethnic nationalism. But (and this brings me to the second development leading to changing perceptions of 'the people') this specification is also being undermined from another direction. This is the growing significance of the notion that the 'logic' of democracy points to the desirability of decision-makers being accountable to all those affected by their decisions. We can call this the 'all-affected' principle. As Whelan (1983: 16) points out, this principle has frequently been supported by recent democratic theorists. May (1978) actually builds this into his definition of democracy, so that 'rule by the people' becomes 'necessary correspondence between governmental acts and the wishes of the persons who are affected by' them. And Dryzek (1999: 44) has recently linked this kind of principle with deliberative democracy: 'an outcome is legitimate to the extent its production has involved authentic deliberation on the part of the people subject to it'.

Now, it is claimed there are crucial difficulties with this idea,[6] although the basic 'difficulty' is precisely that it challenges traditional conceptions of the people. One aspect of this is the way any notion of a people being constituted of varying sets of individuals who happen to be affected by the acts of a particular government departs from the idea of the people as a definite group of individuals. This, however, does not arise where it is the idea of a global people that is being considered. In this case the 'group' being identified is even clearer than in the case of a state-centred conception of the people: 'the whole of humanity' is comprehensive and free of any uncertainties regarding appropriate territorial boundaries. There are questions about the extent to which a people must be a community, and the extent to which there is or could be global community, but these are taken up later. One point that should be remarked on here, though, is the effect of the global warming problem itself. Since global warming potentially seriously affects all the people of the globe, not only does the all-affected principle here directly imply a global people, but also that this people is a community at least in the sense of being a 'community of fate'.

An aspect of the way in which the all-affected principle challenges the traditional conception of the people and causes 'difficulties' is, of course, the way it disregards the state delimitation of that conception. Now, this 'difficulty' would be transcended with the development of a global people, but it is worth dealing with here as establishing the validity of the general idea that a transnational people can pave the way for such a development.

The all-affected principle, then, in violation of the traditional conception, can include as part of the people those affected by a state government's decisions who are beyond the boundaries of that state. The point is that the decisions of the governments of particular – especially the more powerful – states clearly affect persons beyond the boundaries of that state. And this is not just a reference to the field of foreign policy where, by definition, the actions of the government of one state affect other states, and the persons within them.[7] The significance of the point is actually brought out most clearly in the case of domestic policy (or, to put it more appositely, what is conventionally regarded as domestic policy). It is of especial significance that what are seen as the *domestic* actions of a state's government can also crucially affect persons outside that state. These effects can be general and diffuse: for example, actions of the American President and Congress in the field of domestic economic policy can have profound effects on other countries and, indeed, on the global economy. But there can also be effects that are specific and direct. This is notably the case with the transboundary environmental effects of 'domestic' governmental policy – such as a government's policies on power generation, which can cause acid rain in other countries. In all these cases, however, the important point is that the all-affected principle brings in persons *outside* the government's own state, and thereby includes them among those to whom that government should be accountable. And it is this that is seen as the crucial difficulty with the principle: indeed it is seen as an absurdity.

A key reason for seeing an absurdity here is the propensity to presuppose there already exists a self-evidently and uniquely valid delimitation of those to whom a government should be accountable, that is to say 'the people' as traditionally understood.[8] But, of course, precisely what is at issue here is the validity of this traditional notion of 'the people': in particular it is being asked whether it is not a crucial *weakness* of this traditional notion that it no longer captures the realities of who is importantly affected by acts of governments. And I shall argue below that these realities suggest a transnational conception of the people.

These realities are especially salient when it comes to the environment, and it is here that the case for a transnational conception of the people is especially strong. As Doherty and de Geus (1996a: 7) put it:

> [T]he nature of ecological problems suggests the need to consider a redefinition of the form of the democratic community. The impact of pollution may affect those living well beyond the place where it was created which suggests the need to think about democratisation at a transnational level.

Global warming is, of course, one of the – if not the – most serious of ecological problems. Moreover, here the impact of pollution (greenhouse gas emissions) most certainly affects those living 'well beyond' the place of its creation. Indeed, the extent of the impact's effect could not be wider. And, as I shall argue, it is perhaps in respect of global warming that the case for 'democratisation at a transnational level' becomes most compelling.

It should be noted that the above argument about transboundary effects naturally widens out from governmental action to include the effects of impersonal processes (including those causing global warming). There are clearly differences between the two; and I shall comment on these later. But there can also be some overlap: hence the natural widening of the argument. Such overlapping occurs to the extent that, or where, governmental action, and/or the demand for it, results from the nature and impact of impersonal processes. This can be seen in very general terms as a demand for empowerment over our general environment, or 'context': 'there is always a pre-existing context – which affects us – over which we wish to be empowered. To the degree that we obtain such empowerment we establish a state' (Davidson, 2000: 111). Or it can be seen, more specifically, in terms of a demand for governmental control or regulation of particular sets of impersonal processes, such as those causing global warming. The crucial point is that government will be seen as 'allowing' people to be affected by impersonal processes that there should be governmental action to control. And the distinction between being affected by governmental action, and a lack of it, is hazy. (There are, of course, crucial issues here concerning which impersonal processes are in fact controllable, and to what extent, by governmental action. One of these concerns the degree to which certain processes not controllable by single state governments may still in some way be controllable by 'government' or 'governance. I shall take up such issues later.)

An important new idea of democracy emerges from this 'all-affected conception' of the people. A fundamental feature of this idea is the notion of control by an 'extended' people. This involves, on the one hand, extending[9] (i.e. to all those who are affected) the range and number of persons that it is relevant to consider in relation to the acts of a government; and, on the other hand, maintaining that these persons, like the people of orthodox democratic theory, should control[10] that government. An extended 'people' could consist of varying sets of (affected) persons rather than a single entity or single collection of persons. But as already remarked I am interested in a matter that affects all the people of the world, so I am concerned only in that fixed set of people.[11]

A crucial, and innovative feature of the new idea of democracy that can be seen as emerging from the 'all-affected conception' of the people is the notion of the spread of the people beyond state boundaries. Now, this seems to validate control of a government's activities by populations that include those outside the boundaries of the state of which it is the government. Paradoxically, however, a weakening of states (upon which I shall comment in a moment) makes obsolete the very notion of popular power being exercised through popular control of a state; and a 'supra-state' people would need to exercise power in some other way. I shall take this matter up later; but first let me consider the weakening of states by 'globalization'.

Globalization and global democracy

As already indicated, the kinds of change relating to the conception of the people that I have just been considering are closely interconnected with the effect of what are widely seen as important changes in the nature of the international system. The relevant changes are those commonly referred to by the term 'globalization',[12] significant aspects of which are of crucial importance in the global democracy argument.

'Globalization has become one of the central themes in social science in the 1990s. . . . Its widespread use today is clearly connected to the series of major social transformations of the 1990s', including the fall of communism, the end of the cold war, international military operations in the Gulf and the Balkans, the expansion of the World Wide Web, 'and the growing concern about global warming[13] and worldwide ecological crisis' (Delanty, 2000: 81). But what is usually seen as the central process (or set of processes) – and that which most directly concerns me here – is 'the diminishing importance of geographical constraints' (Delanty,

2000: 82), the growing integration of the world economy and the related reduction in the power and autonomy of states, the emergence of global problems and a growing global consciousness.

'Globalization', then, refers to global processes that are said to be changing the nature of the international realm, modifying states' autonomy, interconnecting them more closely together and having the potential to bring about an integrated world society rather than a system of separate autonomous states. The first thing to note is the mutual reinforcement between the effects of globalization and the factors disruptive of the traditional conception of the people I have just been discussing. The central point here is the way in which globalization diminishes[14] state autonomy and sovereignty. Apart from any wider implications, a diminution of the state's importance further weakens the traditional state specification of the people: it undermines the idea that states mark *the* basic divisions of humanity, and thereby constitute the obvious delimitation of separate peoples. But it also undermines the notion that states are *the* agencies for action. Thus what is subverted is both the notion that the state is uniquely the means whereby the people is constituted and also that it is uniquely the means by which it is enabled to act.

Let me focus for a moment on the undermining of the role of states as agencies for action. The argument is that aspects of globalization are diminishing, sometimes drastically, the extent to which states are effective agencies of control. This is usually seen as being most marked in the economic field where it is held that the global economic system is such that states are becoming increasingly enmeshed in, rather than being in control of, events and processes.[15] And of course this has a general impact on the ability of states to act in any sphere. One is the field of the environment. But this in any case displays, in its own way, aspects of globalization – most notably in the case of global warming. In part this is a matter of there being global problems, and – to a degree – a growing global consciousness.[16] Of greater salience, however, is the extent to which this is now a field characterized by supranational processes beyond the control of states, exemplified most importantly (but not only) in the case of global warming. True, states are not themselves really controlled by these processes. But the point is that the processes are not controlled by states: hence (or so it is implied) there is nothing that can control them.

The presupposition is, of course, that such processes should be subject to control. Now, in the economic field there are divergent viewpoints: while many do see globalization as frustrating the necessary regulation of economic affairs, others see it as bringing the benefits of laissez-faire capitalism writ large. But in the field of the environment a predominant view *is* that there should be agencies of control. Environmental protection, it is held, requires this; and there is widespread concern about the way in which globalization undermines the capacity for such control. While some do challenge the view that protection of the environment primarily demands agencies of control, it is accepted here that some such control is essential. To return ro my underlying theme, then, any popular action for environmental protection must entail action through such agencies: hence the concern about the capacity (or, rather, the incapacity) of states for relevant action.

States' lack of capacity for relevant action does not result simply from the diminished power of individual states. True, we do have here another dimension of the incapacity of individual states for action, viz the impotence of *single* states in the face of *global* environmental processes. However, it might be – and often – is argued that the remedy for the impotence of single states is for states to act collectively. Collective action would, by its nature, overcome both the single-state problem as such and, arguably, be a substitute for the diminished power of individual states. But this brings me to another reason for the incapacity of states to protect the global environment: the well-known difficulties in the way of collective action by states. I shall take these up later,[17] but for the moment I need only say that the very structure of the states system militates against cooperative collective action by states.

Globalization, then, disempowers states individually without empowering them collectively.[18] As John Dunn puts it (after suggesting that the threat to the welfare state from the global economy is the central challenge to the viability of the state):

> A less immediate, but potentially more profound, threat to state viability is already beginning to arise from the challenges of environmental degradation. In the sphere of economics the threat to the political standing of the nation state comes essentially from the tension between a national framework of sovereign authority and an uncompromisingly international field of economic causality. In the sphere of ecology, the national framework of sovereign power and governmental responsibility is also in some tension with ecological causality.

> (Acid rain, the ozone layer, still more global warming, are no respecters of boundaries.) But the principal residual obstacle to effective action in the face of ecological hazard . . . is . . . a difficulty in [governments] cooperating effectively. Their problem . . . is not a deficiency in domestic power but a disinclination or incapacity for collective action. (Dunn, 1994a: 13)

There is, then, an incapacity of states to act as agents of control, and especially an incapacity to protect the global environment – most notable in the case of global warming. This has been partly a matter of the erosion of the power of individual states. But it has also been a matter of the inability of states to act collectively. True, this latter is an ongoing feature of the existing international system; but it has been converted into an incapacity for control precisely because of the emergence of global processes that *need* to be controlled. There has been a combination of a diminution of pre-existing power and an expansion of that over which power needs to be exercised, which amounts to a crucial decrease in states' capacity for control.

And, to return to my underlying concern, this means there is a lack of capacity for popular control. In the first place, the incapacity of individual states means they are deficient as agencies of popular control. This in itself, it might be thought, sufficiently demonstrates an absence of popular control, since in orthodox democratic thinking action by the people is conceived only in terms of action through an individual state. Nonetheless, it might still be said that (if it existed) cooperative action by democratic states could achieve popular control. This could be the case where collective control obtained what was desired by the people of each cooperating state. Now, this could not be straightforwardly described as 'control by the people', if 'the people' is conceived as being delimited by a single state. However, the relevance of this point might be arguable in the face of collective control in accordance with the shared wishes of the separate peoples of the cooperating states. But – and this is a crucial 'but' – not only are there the difficulties concerning collective action, there is also a problem about the extent to which separate peoples *do* have shared wishes. There are three relevant possibilities here:[19] (a) separate peoples wanting the same thing, and collective action by their states to obtain it; (b) separate peoples wanting the same thing, and the lack of collective action to obtain it; (c) separate peoples wanting different things. The essential point is that it can be cogently argued that (b) or (c) are more likely to occur than (a) (I shall take up the problem of collective action later). And to the extent that such arguments are valid then there can be no popular control via collective action by states.

There is in fact a significant interconnection between (b) and (c), since a part of what makes collective action by states difficult is often itself an important reason for different peoples wanting different things. That is to say, the disparateness of the wishes of the separate peoples often stems from the very salience and sovereignty of states – and the separate and conflicting interests this produces or accentuates – that stands in the way of collective action. (To be sure, in the case of action against global warming there *is* a common interest and this might translate into shared wishes. This is a crucial point that is taken up below. On the other hand, it may be that precisely what frustrates a proper perception of this common interest and a development of shared wishes is the states system. This, too, is taken up below.) In short, the states system is such that there are unlikely to be wishes shared by different peoples which are translated into action by cooperative action by their various states. Hence, there is unlikely to be popular control via collective state action.

It seems, then, that the nature of the states system is such that the undermining of popular control stemming from globalization is unlikely to be offset by collective action by states to secure the shared demands of separate peoples. (Indeed, as I have just been discussing, that there are such separate peoples – to have separate demands – in the first place, has much to do with the states system.) Hence, in crucial respects democracy, as traditionally understood, is undermined and perhaps even nullified. But it *is* specifically democracy as traditionally understood that is enfeebled. If there is a transnational reconceptualization of 'the people' – as discussed above – and if there are, or could be, processes through which such a people could act, then a reconceptualized and reinvigorated democracy could (and has perhaps already begun to) emerge. It is to a consideration of these possibilities that I shall now turn.

The fundamental problem, then, is that there are events and processes that require popular control but which transcend the ability of states to control them. Accordingly, if there is to be meaningful democracy, then, there must be forms of 'control by the people' that transcend states. This requires both a notion of the people that transcends states, and processes whereby such a people can exert control. I shall consider each of these matters in turn.

Alternative conceptions of 'the people'

I have already looked at the way in which the old state conception of the people is being undermined. And I saw that a crucial aspect of this is the germination of transnational conceptions of the people, fuelled by the all-affected principle and developments in international and global structures and processes. To an extent, the very factors invalidating the old conception of the people are encouraging the emergence of new conceptions. I will need to look at the question of the means by which such newly conceived people might exert control, but let me first look further at the conceptions themselves.

One important question, to which I have already referred, is the extent to which a transnational conception of the people does, or should, incorporate a communal identity. Such an identity is an integral feature of the traditional conception of the people.[20] And arguably this is of considerable importance. Aside from any general communitarian ideas about the importance of community, such an identity converts a mere collection of dissociated individuals into arguably what a people in any notion of democracy must be, a group in which those individuals are supplied with a rationale and motivation for collective action, and a disposition to accept decision-making rules and the demands of social justice. Of course, the kind of communal identity integral to the traditional conception of the people is necessarily missing from a transnational conception of the people. But the question is whether there may be other adequate sources of communal identity.

It is arguable that there are genuine communities that spread beyond states and which are not identified by territorial criteria (Camilleri, 1990; Camilleri and Falk, 1992). Of particular importance in the field of environmental regulation are international 'epistemic communities' (Haas, 1992a, 1992b). Similarly, it is arguable that the community-orientated notion of 'citizenship' – and hence sets of individuals communally linked *as* citizens – can transcend state boundaries and involve multiple political identities and agencies (Hutchings, 1996). And it may be that 'communities of fate' could be generated by the all-affected principle itself. Simply sharing the experience of being affected by, and needing to respond to, the same circumstances, may itself be sufficient to create a relevant communal identity. Although this may be disputed, much would depend on the nature of the circumstances and the experience. And at least modern communications can readily bring affected persons into mutual awareness of each other's existence and their common condition. Moreover, activity in response to this condition can itself involve concrete communal organization and a further strengthening of communal identity. This is well illustrated in important aspects of the nature and activity of the 'new social movements' (Camilleri and Falk, 1992: ch. 8).

However, an important element missing from all these putative communities is the comprehensiveness that characterizes the national/state community. This may be taken as the most important or relevant aspect of the kind of community that, hitherto, has been conceptually linked with 'the people'. Now, there may be room for argument about the necessity of this conceptual linkage, and transnational conceptions of 'the people' of the kind I have just been discussing may be valid and important. Be that as it may, my primary concern here is, of course, with a special kind of transnational conception of 'the people', viz a global people. And this notion of a global people *can* be linked with an idea of community – the global community – that may be regarded as comprehensive. To the extent that such a community does (or may come to) exist it is (or would be) comprehensive at least in the sense that being, like the state, territorially delimited it must in an important sense necessarily include all other types of community within it. Questions concerning the nature and possible existence of a global community will be taken up below. But four points may be noted here. First, the notion of a global community is an established idea; it has, indeed, a long history. Second, there are important mutual reinforcements between the development of a global people and a global community. The former may require the latter, but some of the same factors are involved in the promotion of both. Moreover, the development of a global community may itself be facilitated – may even be dependent upon – the development of a global people and democracy. Third, the very existence of the global warming problem may well be a powerful inducement for the emergence and growth of a global community. Here there is, indeed, a 'community of fate', but on a global scale. Fourth, although territorial delimitation constitutes an important parallel between a global and a state community there are, clearly, crucial differences as well. Besides those relating to the number and diversity of states, there is the difference that a global community would not necessarily be linked with the existence of government. Here the notions of 'governance' – as distinct from 'government'– and of a global civil society come into play instead, but I shall take these matters up later.

The idea, then, of a transnational – and particularly of a global – people can be given some credence. Indeed, as argued above, it may come to have greater credence than the traditional idea of the people. And the grounds for this credibility, as well as the influence of the idea itself, are also grounds for the translation of the idea into reality. But in order to talk of there being a transnational democracy there must also be some account of how a transnational people exercises power. Clearly, given my previous argument, it cannot be by controlling states. What then might it consist in? What could be the nature of rule by a transnational people? I shall now turn to a consideration of this question.

The nature of transnational popular control

In orthodox, state-centred, democracy, popular control – rule by the people – essentially consists in the people controlling the state. The people make decisions about what the state shall do. In other words, there are authority structures (constituting the state) that are used to execute, but do not themselves embody, the popular will. This is a rough and ready distinction and it can be questioned. It does, however, reflect – or is reflected in – the familiar distinction between constitutional rules on the one hand and the enactment of government policy (including legislation) on the other. It is the latter which embodies the popular will. When it comes to our understanding of transnational democracy, however, there is no such distinction, since there is no equivalent of the state. In this case, then, we would envisage the popular will as coming to be embodied in the same set of rules as constitute the authority structures (the nature of which I shall comment on below). This is not, however, the full story since there are institutions forming part of the complex of rules, which are also agents for changing those rules and implementing the policies contained within them. For there to be transnational democracy these, too, would need to be democratized in some sense. There are some complicated issues here, but I shall consider them further under three broad headings: questions concerning the character and democratization of the bodies just referred to, the nature of the popular pressure to which they may be subject and, finally, a characterization of the overall process.

First, then, there is the matter of the bodies[21] that would need to be democratized. The issues here concern their nature and structure, and possible structural reforms to operationalize popular control. The underlying idea is that since, in the model being considered here, these bodies would form a vital part of, and have an essential role in changing and implementing, the complex of rules forming the authority structures in a transnational democracy, they should be controlled by – or at least responsive to – transnational people. The mutually reinforcing purposes of this popular control is for these bodies to become instruments through which transnational people can act and to legitimize, and thereby increase the effectiveness, of their activities. (The question of the nature and function of these bodies will be taken up in a moment.) There is, to an extent, an analogy with the democratic state in the traditional account of democracy, where the familiar mechanisms of liberal democracy, centring on free elections, are intended to give control of the state to the people and to legitimize its activities. The idea is that there should be analogous mechanisms for operationalizing popular control of the relevant international bodies. But since these bodies would not constitute the equivalent of a state, their role – and hence the point of the popular control – needs some clarification.

The essential consideration, which was indeed a fundamental part of one of the lines of argument leading to the idea of transnational democracy, is the extent to which states themselves, singly or collectively, are incapable of regulating supranational processes. The argument then is that it is necessary and possible for such regulation to be undertaken by supra-state institutions and organizations. There are, of course, divergent views among international relations scholars about the feasibility of regulation by such bodies in a situation where the nature of their authority and power is opaque, and where, despite a decrease in the power of states being one reason for such regulation being necessary, the power retained by states and the states system remains crucial.[22] (And, indeed, the term 'international', rather than 'supra-state', is used to describe such bodies, to avoid begging questions concerning the relationship between their authority and that of states.) However, it remains the case that such bodies are widely seen as mechanisms to regulate or control supranational events and processes. There are issues here of central importance. Aspects of these are taken up below; but let me summarize at this point. I start from the position of 'neo-liberal institutionalism', which maintains that in the existing international system some regulation by international organizations is possible (for a useful summary, see the Introduction in Young (1994); see also note 23 below). But I also argue that there is the need and the possibility to extend such regulation further than is presently allowed by the constraints of the states system. And, to the extent that this occurs, democratization of the structure of international organizations would provide for a corresponding measure of popular control of supranational processes.

There are in fact three matters to consider here. First, there is the dual focus of the analysis. On the one hand, there is the issue of the nature of the existing international system and the kinds of initial changes and reforms currently desirable and possible. Arguably these could include some democratization of international institutions, which would help to establish some initial capability for people to have some regulatory influence on supranational processes. On the other hand, there is the issue of the nature of the genuine transnational democracy which would transcend the present international system and where the former international – and now global – institutions would have greater power, and their democratized structure would mean a greater measure of popular control of what were formerly transnational processes. But, of course, the initial democratization would itself promote the development of genuine transnational democracy. The other two matters concern the nature of the authority and power of these international-cum-global institutions, and their role in making changes in the complex of rules that I see as constituting transnational democracy; but I shall take these up later.

Let me now focus on the existing international system. Within this there is a variety of bodies, a purpose of which is to regulate supranational events and processes. And the question of their democratization is already becoming a matter of growing concern. For example, in a speech at the South Summit in 2000 Thabo Mbeki said: 'We have to ensure the democratisation of the international institutions of governance, including the UN and the Bretton Woods institutions' (reported in the *Independent*, 2 May 2000). A key issue here is the need for legitimacy. For example, Andrew

Marshall wrote in the *Independent* (13 March 2001) of 'The IMF's struggle to gain legitimacy', but generalized the central theme:

> These are not problems of the IMF itself – they are problems of the infrastructure of global governance erected over the past 50 years. . . . {O}rganizations that could once be sustained through intergovernmental agreement, diplomatic exchanges between ambassadors behind closed doors, are now struggling for a broader legitimacy and failing to find it.

At the moment international bodies can be seen as the creatures of states: they are overwhelmingly set up by, their personnel are appointed by, and they are accountable (if at all), to states. There are, moreover, numerous international regimes[23] which undertake regulation. These may involve more than the operation of one body, but to avoid circumlocution I shall refer to them simply as bodies. I am primarily interested in those which are global, though, as before, it is worth being alert to all possibilities for transnational democracy as paving the way for global democracy.

At present international institutions or organizations are, then, typically controlled by, or accountable to, assemblies of states. So notions of 'democratic' control can get no further than control by an assembly in which relevant states (i.e. those which are affected by the decisions of the international body concerned) are equally represented. Not that this is wholly unimportant. Equality of representation of all affected states *is* an important issue. Frequently, one of the allegations included in condemnations of international bodies as undemocratic is that the interests of the rich states of the North, as compared to the poor states of the South, are given undue weight; and lack of – or inadequate – representation of the latter may be one manifestation of this (see further below). However, even equal representation of affected states is unsatisfactory; and for three main reasons. First, states can have populations of very different sizes; and equal representation of states may well mean unequal representation of people. (There is also, of course, an argument linked with the very processes of globalization that helped raise the aspiration for transnational democracy in the first place. According to this, even equal representation of states of equal population size would not necessarily mean equal representation of people's interests, since states are subservient to economic forces that serve the interests of richer countries. This issue is taken up below.) Second, representation of states usually means representation of governments rather than peoples. This is obvious in the case of states which are clearly non-democratic; but even in otherwise democratic states foreign affairs are seldom under meaningful democratic control. Third, even in so far as peoples *are* represented through states it can be cogently argued that this is to miss two crucial points. On the one hand, such representation may be 'distorted'. It is in their capacity, and in terms of their interests, as members of a particular state that people are being represented, whereas it may well be other capacities and different interests that are most relevant. This point is especially important in the case of global warming, and will be taken up below. On the other hand, and for the reasons already considered, it is arguable that members of a state do not, in any case, constitute a relevant 'people'.

However, as suggested above, instead of the present system there could develop the notion of control by, or at least, accountability to, transnational peoples. 'International institutions and organizations' (or, rather, the successors to what are now called such[24]) could then become the instruments of popular power and thus important features of transnational democracy. Although what could be seen as anticipations of this are already developing,[25] at this point it is difficult to say what the processes of such control or accountability would amount to. However, it is worth noting some suggestions. In a slightly different context Held mentions 'the possibility of general referenda cutting across nations and nation-states . . . with constituencies defined according to the nature and scope of disputed problems' (Held, 1995: 273). And Brown, after invoking the principle that 'those whose behaviour substantially affects the well-being of others are accountable to the affected', argues:

> At a minimum, the principle [applied in the international realm] would require that potentially victimized populations at least be consulted by those whose actions might cause major harm. At a maximum, the substantially affected would be provided with a veto over actions they did not like. The international procedural and structural embodiments of such accountability obligations are potentially far-reaching, highly controversial, and will be extraordinarily difficult to negotiate in some fields, especially where the affected populations demand not merely a fair opportunity to express their view but also seats and voting power at the principal tables of decision. (Brown, 1988: 306[26])

Held's remarks are in the context of his general model of 'cosmopolitan democracy'. But he does say something about international organizations in particular:

> [T]he opening of international governmental organizations to public scrutiny and the democratization of international 'functional' bodies (on the basis perhaps of the creation of the elected supervisory boards which are in part statistically representative of their constituencies) would be significant. Extensive use of referenda, and the establishment of the democratic accountability of international organizations, would involve citizens in issues which profoundly affect them. (Held, 1995: 273)[27]

The idea, then, that international bodies should be accountable to transnational peoples is beginning to be discussed. There are, it is true, also sceptical voices. One is Robert Dahl's (1999). Dahl has three main arguments. First, large organizations are difficult to control democratically. (An aspect of this is that – as in respect of the foreign affairs of traditional states – popular knowledge will be inadequate (Dahl, 1999: 23–4). But I have already addressed this issue, in Chapter 2.) Clearly this is an important argument. But historically the main difficulty was expanding democracy from the Greek *polis* to modern states as large as the United States, and it is unclear whether there are comparable difficulties in expanding it further from the scale of the United States to international organizations. Second, there is the argument that in respect of international organizations disagreement is inevitable 'among a large group of persons', and 'civic virtue [will be] too weak a force to override individual and group interests' (Dahl, 1999: 26). Again, this is important; and I shall be addressing aspects of it below. But it does seem to assume that supranational peoples are not communities – an assumption that I have already disputed. Dahl's third main argument is that existing international organizations 'such as the United Nations are substantially immune from even the limited popular control that is characteristic of foreign policy in national democracies' (Shapiro and Hacker-Cordón, 1999a: 4). However, this is to beg the question since the issue I am now considering is, precisely, whether and how international bodies can be subject to popular control.

To take this consideration further, let me now turn more specifically to the type of international organizations in which I am primarily interested, viz those that are global in scope. It is noteworthy that the organization mentioned above by Dahl is the United Nations. And it is above all the UN itself upon which discussions of global democracy tend to focus. I shall return to this in a moment, but there are also subsidiary and other institutions and organizations that are important and global in scope. Some of these, notably the World Bank, the IMF and the WTO, may be seen as having a particular rather than a general remit. But activity and regulation in the fields of finance, economics and trade have such wide implications that there can be doubt about the applicability of this distinction. For example, the decisions of the WTO relating to free trade have wide-ranging implications. Indeed, this body has attracted the ire of environmentalists for the adverse environmental impacts of its decisions. The question of democratizing the WTO has also become important since being highlighted by the Seattle riots of December 1999 (see Andrew Marshall in the *Independent*, 13 December 1999). There has been a similar development with regard to another body that does, now, have a more overtly general remit – the annual summit of world leaders known as the G8 (formerly the G7). This may not be an 'organization' as such, but with the structured regularity of the meetings[28] it amounts to an 'institution'. Moreover, it is generally conceived as being very important, perhaps the only global body that could be described as powerful. It is also, clearly, 'unrepresentative': 'The G7 [as it then was] represents only 12 per cent of the world's population. By excluding China and India, it can no longer even claim to represent the world's major economies' (The Commission on Global Governance, 1995: 154). But, as this quote from *Our Global Neighbourhood* suggests, this very unrepresentativeness is leading to demands for, and the beginnings of, the 'democratization' of the G8. This was especially evident in the wake of the 2001 Genoa summit. The *Independent* (20 July 2001), for example, asserted that: 'Contrary to the claim being made by many of the protestors besieging today's G8 summit in Genoa, this annual circus is beginning to be a tentative but real example of global democracy in action.' This has resulted from a combination of an expansion of membership and remit, and growing pressure from global public opinion. It is worth quoting further, at some length, from the *Independent*'s editorial:

> It is true that a scripted meeting of eight men in suits representing the richest nations in the world (and Russia) . . . does seem an unlikely forum in which to acknowledge the

> interests of all the six billion mostly poor people on the planet.
>
> Yet that is precisely what is happening, however imperfectly. The summits began, as the G3, as an informal attempt to co-ordinate [economic management]. Increasingly formalised and institutionalised . . . the G7 broadened its agenda . . .
>
> Since then, the summits became the focus not just for anarcho-hooligans but for principled protest on behalf of the world's poor . . .
>
> It is apparent from this year's summit that this modern form of popular pressure has yielded its rewards [in respect of moves on global poverty and Aids].

And, with direct relevance for our concern with global democracy, the editorial goes on to assert:

> Although the future of the Kyoto Protocol is not on the formal agenda . . . it too is bound to be raised on this highly visible public platform. President Bush has been forced by international opinion to declare before arriving at the summit that 'we share the goals of reducing greenhouse gases'.

As we can see, in the case of the G8 the focus is, in fact, more on the nature of popular pressure, which comes under the second of the headings enumerated above. Indeed, it is unclear what relevant structural reforms there could be:[29] rather, any exercise of popular power is a matter of popular pressure on a body that is not itself democratic. However, when we turn to what is usually seen as the key body in any discussion of global democracy, the United Nations, the reverse is the case. When it comes to developing an organization as a mechanism for control by a global people the UN is the obvious candidate. And such development would be a matter of reforming the UN's structure. It should be noted that the UN has already been seen as 'the most significant global institution embodying the democratic ideals and aspirations associated with the new world order' (Gershman, 1993: 5). This, however, is primarily a statement about the global promotion of traditional democracy, i.e. within states, rather than the promotion of global democracy. But the UN is also now coming to be seen in terms of the latter. Now, the UN currently exists in a world of states; and there is a tendency to see it becoming an instrument of 'world democracy' in the sense of being – with suitable reform – an instrument of all the world's states, equitably represented.[30] The former Secretary General, Boutros Boutros-Ghali was a powerful advocate of this (Boutros-Ghali, 1996). But true global democracy requires an organization that is directly an instrument of a global *people*. Representation of states at present only amounts to indirect representation of (traditional) peoples, and de facto amounts to representation of governments. Of course, a possible reform would be to have delegates[31] to the UN directly elected by, and accountable to, the people of each state. Held advocates this (see note 30), as do Barnaby (1991), Archibugi (1995), Bienen, Rittberger and Wagner (1998) and Archibugi, Balduini and Donati (2000). This could overcome the problem of the disparity in sizes of states since – as in the case of the American House of Representatives in contradistinction to the Senate – delegates could be apportioned in relation to size of population. But it would not get round (though it might conceivably diminish – see below) the problem, noted above, of the crucial differences between a 'world aggregate' of state-defined peoples and a global people. Galtung (2000) has some interesting thoughts about building institutions to reflect and induce a global focus, but even he has state-based constituencies for a United Nations Peoples' Assembly (though he considers it a mistake to think it is only the United Nations that has to be democratized, and his proposals here are but part of a radical overall scheme of ideas for global democracy).

To be fully an instrument of global democracy, then, the UN would need to be transformed into a body directly representing the global people. This would, indeed, be a transformation since the UN is inherently a part of, and is structured as a body representing, a world of states. And it would be more realistic to focus initially on the idea of a UN run by elected state-based representatives. Unmodified global democracy remains the desired end – and, indeed, for my overall argument transcendence of the states system is a crucial feature of global democracy – but there need to be realistic first steps. And it is worth noting that Archibugi, Balduini and Donati (2000: 136) say of this half-way house that 'representatives would be elected directly by citizens and would therefore be more keen on promoting global policies rather than state-centred ones'.

As I shall argue below, global democracy generally would facilitate action against global warming. But there are some global structures and processes that are of particular relevance – those specifically concerned with the global warming problem. Since they can also be seen as potential structures of global democracy they clearly have particular importance in considering global democracy's contribution to combating global warming. (And, indeed – as another example of reciprocal interaction – it can be seen here that the need for structures to bring about action against global warming has possibly provided potential structures of global democracy.) The structures in question are mainly to do with the developing international regime, or set of regimes, deriving from the 'Rio process' – the 1992 'Earth Summit' and the ensuing follow-up conferences, most importantly those at Kyoto in 1997 and Bonn in 2001 which produced and revised the Kyoto Protocol.

This can all be seen as an offshoot of the UN: the Earth Summit's official title was 'The United Nations Conference on Environment and Development' and it 'reported ultimately to the political authority of the UN General Assembly' (Grubb *et al.*, 1993: 18).[32] But, clearly, this amounts to a separate set of global environmental regimes, with the potential to become very important indeed. In what ways, then, might they be, or become, democratic? First, there are the conferences themselves. These are 'democratic' in the sense that all participating states are equally represented. Since it is states that are being represented this is, as argued above, only a bogus or partial sense of 'democratic'. Further democratization would involve the people of each state electing their delegates to the conferences. And, again, while full global democracy would require election of delegates by the global people, direct election by state peoples could be a realistic first step. Second, there are the mechanisms set up – or restructured – to implement the decisions of the conferences, such as the Commission on Sustainable Development (CSD) and the Global Environmental Facility (GEF). With reference to the latter case, The Earth Summit's Climate Convention of 1992 established a mechanism for the management of financial transfers (between rich and poor nations – see below). This was to be operated by the GEF, and the question of whether this body was sufficiently democratic subsequently arose: 'During its renegotiation in March 1994, the GEF was restructured to make it more democratic. Indeed, a Governing Council with equal representatives from North and South[33] was established' (Rowlands, 1995: 211). And:

> By rejecting sole use of the 'one dollar, one vote' system of the international financial institutions, recipient countries were given some say in the final decisions of the Facility. More specifically, the Facility will have a Governing Council, which will have two voting methods . . . The first is based on money: the four largest donors . . . will . . . predominate. The second is based on membership of the council. (Rowlands, 1995: 204)

Again the question of democratic structure is seen to turn on the problematic notion of representing states. As before, this can be seen as a realistic first step; but, apart from this, the point is to illustrate that – and how – the issue of democratizing a global organization concerned with global warming is already arising. And in respect of the CSD the notion of simply representing particular states is in any case arguably diluted because of its status as a UN body: it is a functional commission of, and its members are elected by, the UN Economic and Social Council from among the member states of the UN.

Let me now move on to the second of the headings specified above. This referred to processes, outside of the organizations themselves, by which transnational people seek to influence or control international organizations – i.e. the concern here is with the transnational popular pressure on, rather than with the structure of, those organizations. As a lead into considering the nature of this popular pressure we should focus on the point that actually treating it as significant does of course challenge important aspects – or a dominant variety – of the globalization thesis. That is to say, to see organs of international governance as subject to popular pressure is to reject a view of the nature of global processes and actors that sees power lying with the forces and agents of global capitalism. This is a view that brushes aside the argument that states' loss of power has been remedied by a transfer of regulatory functions to those organs, and maintains instead that they too are agents of global capitalism. While this is held to be partly a matter of the undemocratic structure of international organizations, the more fundamental proposition is that such organizations are in any case the victims or creatures of the underlying undemocratic nature of the global system they inhabit. This is the kind of view that was manifested, for example, in attacks on the WTO at its 1999 Seattle meeting. As Andrew Gumbel put it (*Independent*, 29 November 1999): 'While the WTO would like to paint itself as the instrument of greater global affluence, its detractors believe it poses a grave threat to democratic accountability by only representing the interests of multinational corporations.'

The challenge, however, to the features of the global order portrayed by this view is as significant as those features themselves. Not only do these features provoke a backlash, but, as I have argued above, some of the elements of the complex phenomenon of globalization themselves contribute to global democratization. Global democratization can be seen both as a remedy for, and as promoted by, globalization. This ambivalence is captured by Richard Falk in his account of 'globalization-from-above' and 'globalization-from-below' (Falk, 1999, 2000). The processes of globalization, then, do not rule out the notion that organs of international governance can be subject to popular pressure; far from it.

So let me now turn to the nature of this popular pressure. Very roughly speaking I could say there are two ways of conceiving – or perhaps that there are two forms of – transnational popular pressure. On the one hand, there are unmediated manifestations of international or global public opinion, examples of which I have already referred to. On the other hand, there is group activity, which I shall now consider. Such activity may be understood simply and in itself as constituting a form of popular pressure; or it may be that transnational popular pressure is better understood in terms of the operation of transnational civil society, of which this group activity is a key component. The groups concerned are usually referred to as international non-governmental organizations (INGOs) or more simply as non-governmental organizations (NGOs).

So let me now consider the role of INGOs. To begin with the first of the above understandings, just as within states, pressure groups can be important influences on governmental action, so in the international or global realm INGOs are becoming increasingly important influences on the actions of international organizations. The World Bank provides an example: 'From environmental policy to debt relief, NGOs are at the centre of World Bank policy. Often they determine it' (*The Economist*, 1999: 23). It is, in fact, in relation to environmental affairs that INGOs are especially important. This is particularly true with regard to the problem of global warming: 'The watershed was the Earth Summit in Rio de Janeiro in 1992, when the NGOs roused enough public pressure to push through agreements on controlling greenhouse gases' (*The Economist*, 1999: 23). Indeed, INGOs now normally have an important presence at international conferences: 'Perhaps the most telling indicator of NGOs' prominence in world politics is their increasing presence in international conferences' (Princen and Finger, 1994: 4). This first became important at the Earth Summit (Princen and Finger, 1994: 4–5; Grubb *et al.*, 1993: 11–12), but is now widespread (Princen and Finger, 1994: 5–6). But it is not just at conferences that INGOs now have an officially recognized presence; they have also become involved in the running of international organs of governance or the implementation of their decisions. Again, it was the Earth Summit that was pivotal. 'The UNCED instruments . . . strongly endorse the role of non-governmental organizations in the activities of international institutions and in the elaboration of treaties in the field of sustainable development' (Sands, 1998: 90); and there was 'particular emphasis on the role of "major groups" in implementing Agenda 21'[34] (Imber, 1997: 220). But it is not only in relation to the 'Rio process' that there is this pattern of INGO involvement: a similar pattern is now important in relation to a range of international bodies in the field of the environment (Princen and Finger, 1994: 5).

Now, within states pressure-group activity can be seen as providing for popular participation and influence; this is a widely held view and was held to be a key feature of democracy in what I might call orthodox mid-twentieth-century 'pluralist democratic theory'[35] (see, for example, Holden, 1993: 109–11). So, too, it can be held that the activities of INGOs, by providing for popular participation and influence, may become a fundamental feature of global democracy. (This way of putting it assumes that an INGO, like a pressure group, is a group within, rather than itself constituting, a people: see further below.) There are two, overlapping, ideas here. First, there is the general contention of 'orthodox pluralist democratic theory' that group activity provides for participation by the people. Second, there is the additional notion that INGOs supplement – or, rather, remedy the defects of – the states system. As Mark Imber (1997: 201) puts it:

> [A]lthough citizens can participate *indirectly* in processes of geo-governance, through the election of governments which conduct [the relevant] multilateral diplomacy, can citizens participate *directly*? The enormous expansion of Non-Governmental Organizations (NGOs), such as Amnesty International or Friends of the Earth, and their international affiliates and

> confederations (INGOs), does now allow this to occur [emphasis in the original].

And he drives the point home by adding that the role of INGOs 'in geo-governance is a form of participatory, if not direct, democracy' (Imber, 1997: 210).

But these ways in which the activities of INGOs are seen as an essential element of transnational democracy, and which parallel those portrayed in pluralist democratic theory, are supplemented by two others. The first concerns the extent to which INGOs, supplementing the role of international organizations, provide vehicles for the embodiment of the popular will that would not otherwise exist. In orthodox pluralist democratic theory, typically, the popular will – articulated and pressed on government by pressure groups – is for modification of ongoing governmental action. In the international sphere, by contrast, there is no 'governmental' action as such, and there is often a lack of any analogue to it, so it may only be INGOs that can initiate – or even implement – action that is popularly desired. This is especially true in relation to environmental matters. Here INGOs have a vital role: 'perhaps the most noticeable change in the conduct of international relations visible in the environmental arena is the central role of NGOs' (Thomas, 1992: 18). But, more than this, their action is 'absolutely essential to most environmental action . . .' (Princen and Finger, 1994: ix, quoting Caldwell (1988: 24). In relation to global warming in particular, 'NGOs will be key actors in the process of defining appropriate policies and facilitating their implementation' (Oppenheimer, 1990: 340). The second, and crucial, additional way in which INGOs provide an important element of transnational democracy is that by transcending states they provide expression for a public opinion that is transnational in origin and/or outlook (this may, indeed, be one reason for their prominence in the environmental arena). This is of especial importance in relation to global warming and will be commented on below.

Now, it might be argued that a process of influence by INGOs is in fact *un*democratic. 'Some will celebrate [the INGO process] as the advent of the age when huge institutions will heed the voice of Everyman. Others will complain that self-appointed advocates have gained too much influence' (*The Economist*, 1999: 24). And the same article (23), while noting that NGOs are 'increasingly powerful at the corporate, national and international level', asks whether 'they represent a dangerous shift of power to unelected and unaccountable special interest groups?' In part, then, the question arises because INGOs are unelected and unaccountable: 'NGOs may *claim* to constitute "we the peoples" . . . but as membership subscription organizations and not *elected* representatives, their claim to be "more representative" than elected governments is contestable' (Imber, 1997: 210; emphasis in the original). (But note that the latter aspect of the criticism ignores the point about providing for transnational public opinion.) INGOs, moreover, may not be internally democratic. In short, 'NGOs are not part of any formal democratic process. [And] most NGOs are far from democratic in their internal practices and procedures' (Evans, 1997: 133). Indeed, 'as unelected, activist-driven organizations these NGOs may in fact institutionalise an elitist, rather than a participatory, tradition' (Imber, 1997: 220).

I will question this line of argument below, but there is also the further, but associated, point about 'special interests'. The underlying argument here is that the INGO process involves unbalanced 'representation': that INGOs promote particular interests rather than the views of transnational people in general. At one level this may be a matter of particular economic interests versus the general good – which, from our perspective, is that which is environmentally good. The comparison with the intra-state pressure group process suggests that organizations concerned with economic interests are a legitimate and crucial component of the global group process. But to count such organizations in is to risk economic power negating the general good – and arguably therefore the 'popular will' – while to count them out may amount to undemocratic exclusion. The influence of oil companies, for example, in relation to global warming negotiations is widely seen as sinister, but would it not be undemocratic to attempt to exclude them? There are some parallels to issues raised by the contention of critics of orthodox pluralist democratic theory that the participation provided by the pressure group process is crucially distorted and less than comprehensive (Ricci, 1971). Now, although this contention can be challenged (Kelso, 1978), it does raise genuine issues about the undue influence of economic interests. More important, though, are the differences between the processes covered by orthodox pluralist democratic theory and those being considered here. The point is that with regard to the former, I am concerned with all the groups impinging on policy-making by the state, and here *interest* groups are dominant.

In the case of the latter, however, I am specifically concerned with *cause* groups.[36] A fundamental reason for this is the tendency for it to be 'causes' rather than interests that unite people beyond state boundaries. (This has important connections with the point, developed below, that it is global democracy that can focus public opinion on action against global warming, rather than state democracy which structures public opinion towards the narrow particular interests of individual states.) An important aspect of cause rather than interest groups being central is that problems concerning 'selfish' wills giving rise to 'the will of all' instead of a 'general will' (see, for example, Holden, 1974: 134–5) do not arise. (Indeed, an INGO might itself have a 'general will' since, unlike a pressure group, it might in some sense be seen *as*, rather than a group *within*, a people: see below). But the central point is that the problem of the undue influence of economic interests does not arise. The work of INGOs is *distinguished from* that of 'the profit-making' sector (Princen and Finger, 1994: 11–12). It is not that interest groups are unimportant influences on the organs of international governance; far from it. Rather – as we have already seen – it is that the processes of transnational democracy are conceived as being *opposed* to the influence of economic interests. This is 'globalization-from-below' as opposed to global capitalism and 'globalization-from-above': 'Here is a democratic response to those forces of globalization typified until now by the rise of transnational corporations [and] deregulated markets' (Imber, 1997: 220). To put the point another way, the traditional concept of democracy is such that its application must cover the whole process relating to the state and the formation of its policies, the operations of interest groups included. (Of course, this does not itself dispose of the different question of whether, or when, such an application does in fact show that the process *is* democratic.) The concept of transnational democracy, on the other hand, is not anchored in the state and its very application in the first place is determined by picking out elements (actual or potential) of popular power.[37]

The INGO process, then, is not dominated by economic interests. However, in another way it may be 'unbalanced' by the domination of INGOs from the North. As Rowlands (1995: 249) points out, Northern INGOs 'have traditionally been concerned with the "environment", while [Southern INGOs] have been more interested in "development"'. (For a discussion of Southern NGOs, and their relationships with Northern NGOs, in the field of development, see Hudock (1999).) The Northern NGOs dominate in environmental matters – not least in respect of global warming – and citizens of Southern countries see some of them as exerting great power. 'Consequently, there is a feeling that the large international NGOs' groups from the North (many of which have dominated international meetings on issues of global atmospheric change) may not necessarily represent broader public opinion' (Rowlands, 1995: 249–50). On the face of it this is an important line of argument. However, it misses two important points. First, INGOs specifically do not 'represent' territorially. They promote causes rather than represent the interests of territories, or those within them. As in the case of the promotion of environmental causes, this should be of universal benefit. A central 'role for NGOs' it has been said, is that of 'representing . . . humanity at large' (Oppenheimer, 1990: 342). Second, because of this INGOs speak on behalf of all people, and where there is conflict with dominant perceptions of local interests it may be that local people *can* only be represented by non-local INGOs. Global warming is a prime case in point. To the extent that people in Southern countries want action against global warming they must look to 'Northern' INGOs, since their governments and 'their' INGOs give priority to development issues. It might well be objected that this priority does in fact reflect what Southern people want. But here a point made in Chapter 2 about the implications of our basic premise is relevant. The premise is that combating global warming *is* desirable; and the point concerned its role in the validation of the prioritization of long-term over short-term interests. Here, we can say that it follows from the basic premise that action against global warming *would* benefit Southern people; and hence that the democratic stance concerning ordinary people's knowledge implies that such action is what Southern people will want. Now, if indeed this is *not* what they want, then my general argument about global democracy promoting action against global warming is undermined. But judging whether this is so is a complex matter. Two points must be borne in mind. First, there are complex connections between development and global warming issues. Arguably, what Southern people want is action against global warming, provided that policies for such action are socially just and take account of development issues. And this is just what many Northern environmental INGOs campaign for. (Fundamental questions here, concerning international social justice and policies for combating global warming, will be taken up below.) The second point concerns the potential the states system has for distorting the perception and formation of public opinion. The potential effect on public opinion of the distorting lens of national interest has been remarked on before. But there can also be difficulties in distinguishing governmental and public opinion, especially in the case of non-democratic states; and among Southern states there is a significant number of these. Thus even if state-centred expressions of Southern views may suggest an antipathy to causes embraced by environmental INGOs, it can be argued that the latter articulate a more authentic representation of Southern public opinion.

The INGO process, then, is not dominated by special interests, economic or regional. But what about the criticism that they are unelected, unaccountable and elitist? This is partially to be answered by referring to tenets of orthodox pluralist democratic theory where the group process is held to provide for participation by all the people, and to do so – at least in some ways – rather better than the electoral process (Holden, 1993: 109–10; see also Kelso, 1978). This may not be a full answer, since the group process of orthodox pluralist democratic theory is a supplement to, rather than a replacement for, a framework of governmental elections. But two points should be made here. First, it may be held that in a full theory of global democracy there would be a better parallel with orthodox pluralist democratic theory, and the INGO process would operate within a framework of elected organs of international governance. Second, if, or to the extent that, the organs of international governance do not become subject to popular election, then the INGO process, whatever its faults, is left as *the* central democratic mechanism here: in the absence of elections the operations of INGOs must be the critical mechanism for transmitting transnational public opinion to those organs of governance, and holding them to account. And, it should be remembered, this means of course that INGOs are vital to the role of democracy in achieving a successful response to global warming: 'the determination and implementation of policies for responding to climate change will be successful only with broad public involvement in decisions; and the NGOs are a mechanism particularly suited for bringing about such involvement' (Oppenheimer, 1990: 345).

In the above ways the operations of INGOs can be seen as constituting a central process of transnational democracy rather in the way that orthodox pluralist democratic theory sees the pressure-group process as constituting a central process of intra-state democracy. But there is also another view. According to this INGOs are not, like pressure groups, simply groups within a people, articulating its will. Rather, they may also help *constitute* a people. I have already discussed the extent to which transnational conceptions of the people involve the idea of transnational communities, and INGOs can be seen as playing a vital role in generating such communities. INGOs can be regarded as manifestations of 'social movements', on an international scale (Finger, 1994; Della Porter and Kriesi, 1999) and thus as transnational communities: 'social movements are, fundamentally, political communities. . . . They [lead people] to re-conceive their own identities. . . . They have no necessary bounds in space and time' (Magnusson, 1990: 54). Such communities, then, can be seen as constitutive of a transnational people or peoples.[38] This could be linked with the notion, referred to earlier, of there being various transnational peoples, so that an individual community might be regarded as constituting a people. But the more intelligible idea, and the one which interests me, is that of the sum total of such communities providing the structure for a global people by constituting a global civil society.

The idea of a global civil society has lately become important. Moreover, as suggested by the extent to which it is environmental groups that constitute the growing number of important INGOs, the development of global civil society can be seen as especially connected with the growth of environmental concerns (Lipschutz, 1996). The idea itself involves transposing the domestic concept of civil society to the global arena. Referring, for example, to Cohen and Arato (1992) and Walzer (1992, 1995), Lipschutz (1996: 2) characterizes the 'notion of "civil society" [as having] a long history, but it generally refers to those forms of association among individuals that are explicitly not part of the public, state apparatus, the private household realm, or the atomistic market'.[39] Analogously, in the global arena INGOs, being associations which are not part of state apparatuses or the global market, are usually seen as the essential elements of global civil society. Falk (2000: 162), for instance, refers to NGOs as 'those actors associated with global civil society' and Dryzek, talking of the 'role of transnational civil society in environmental affairs', says that 'the most prominent actors here are non-governmental organizations (NGOs) and what Wapner calls TEAGs, "transnational environmental activist groups"' Dryzek (1999: 44, referring to Wapner, 1996), while

Lipschutz (1996: 77) conceptualizes the influence of NGOs in relation to the 'Earth Summit as well as a host of other international conventions' as the influence of global civil society.

INGOs, then, and the structures and processes of their operations, can be seen as constituting (at least crucial elements of[40]) a global civil society. This means that on the one hand the prominence of INGOs is taken as demonstrating the existence of a global civil society, while on the other hand another and significant dimension is ascribed to their role. But what exactly is the import of the idea of there being a global civil society? It is important on two levels – as a proposition about what structure the global realm has (but see note 46) and as an interpretation of the meaning and significance of that structure.

The proposition about the structure of the global realm has two central distinctive features. First, there is the crucial modification, or perhaps rejection, of the state-centric view of an 'international' realm. Instead a *global social* system is postulated which is independent of – and perhaps more important than – the states system: global civil society like domestic civil society consists of bodies functioning independently of state authority. Secondly, again as an analogue to domestic civil society, this is a social system distinct from, and perhaps in opposition to, the market system:[41] as suggested in the earlier reference to 'globalization-from-below', there is a contrast between global civil society and global capitalism.

The interpretations of the meaning and significance of this global social system that concern us here relate to implications concerning global democracy. This is both a matter of transposing understandings of the importance of domestic civil society for orthodox democracy (within the state) onto the global realm, and of seeing global civil society as generating necessary conditions for the development of global democracy.

Taking up the former point first, it could be said that, broadly speaking, there are two forms of connection between democracy and civil society. First, analyses of civil society can add substance to accounts of 'the people'. They give sociological content and grounding to notions of a people by providing a sophisticated account of the nature and possibility of a pluralized society which is nonetheless a community. In a similar vein, the existence of civil society can be viewed as a necessary condition for the existence of democracy.[42] Second, and overlapping into matters discussed under the third of my headings – understanding of transnational popular control – which are taken up below, civil society can itself be seen as the realm of democracy. This notion became prominent in the 1980s as revived theories of civil society were used in relation to the growing democratic opposition to the totalitarian states of Eastern Europe. Such theories were used to show such opposition as taking place within, and as involving the nature of, civil society (Cohen and Arato, 1992). Civil society – in contradistinction to the state – is seen as a realm of freedom and spontaneous self-government. There are resonances here with other 'non-governmental' understandings of democracy,[43] though with their theories of social integration giving an understanding of a pluralized *community*, civil society theories have their own account of non-governmental societal ordering. Buttressing the view of civil society as the realm of democracy is the notion of civil society being conducive to deliberative democracy, since it is seen as 'a realm of relatively (though of course not perfectly) unconstrained communication' (Dryzek, 1999: 45).

Domestic civil society, then, has been seen as crucial for state democracy. And this is a reason for analogously regarding global civil society as important for global democracy. But the specifically global character of global civil society has its own additional significance. This is principally because of the additional importance at the global level of the underpinning of the notion of a people by the theorizing of a diversified community.[44] There are four important, and interrelated, points here. First, there is a general point about the connection between the notion of a global people and the development of a sense of global community and global consciousness, and the role that global civil society has in this development (Lipschutz, 1996: 52). Associated with this – and of particular relevance here – it can be said that the link with environmentalism means that the 'perspectives of humanity' are promoted by global civil society (Falk, 1995: 169). Second, the theorizing of a diversified community has a particular significance for developing a global consciousness in the face of the global dimension of diversity. Third, in contrast to what obtains at the state level, the credibility of a notion of a people (a global people) is to be established rather than assumed. Fourth, the model of a 'self-ordering' community has particular importance at the global level, where there is, notoriously, an absence of ordering by an overall polity (Lipschutz, 1996: 52ff).

This brings me to the third of the headings for organizing my discussion of transnational popular control – a consideration of the general nature of the postulated processes of control. An important question here concerns the extent to which the central conception is analogous to one in which a popular will is implemented through the action of government, a conception which involves regarding popular power as a matter of controlling or influencing the operations of government. Although there are some exceptions (see note 43), this is, after all, the typical conception in orthodox, state-centred, theories of democracy.

The answer to this question is complicated. One of the complications concerns the spread in the meaning of the term 'transnational'. There is both the global and the non-global connotation. As I have already observed, although my focus is primarily on the former, shared features mean we also have an interest in the latter, from which it cannot always easily be disentangled. Now, in respect of this latter, the influence of the 'orthodox conception' of popular power is quite strong: as was suggested earlier, there is a temptation to see the idea of the control of an international organization or regime by a (non-global) transnational people as an analogue of the notion of the control of a government by the people of a state. However, where 'transnational' is understood as 'global', additional considerations come into play. This is for two main reasons. First, in the non-global case there really is only the dichotomy of (an analogue to) 'the government', on the one hand, and the (transnational) people on the other, so that it is difficult to conceive popular power other than in terms of the control, or influence, of the latter over the former. There is no equivalent to the notion of a self-governing entity that is provided by the idea of a global civil society (see below). Secondly, in the global case there are the familiar difficulties with the very idea of there being a global analogue to the government of a state. The idea of there being a world government and polity is widely seen as an aspiration that is either utopian or dangerous. And while it is true that some conceptions of global popular control involve the notion of a global people controlling a global government – usually conceived in terms of a revamped UN – these are widely regarded as naive. More sophisticated conceptions involve more nuanced notions of the exercise of global popular power. Here I come back to the idea of a global civil society. But there is also a connection – somewhat subtle and complex – between this and another important idea, that of a distinction between 'government' and 'governance'.[45]

I have already remarked on the notion that civil society, being a self-ordering community, is the realm of democracy; and, according to this idea, global democracy would be constituted by a global civil society.[46] One interpretation seems to be that civil society – hence global civil society – is a realm of spontaneous self-government. However, it is not too clear just what this means, and in practice use of this idea that global democracy is constituted by global civil society tends to involve some analogue of the idea of government, treated either as an element of, or as subservient to, global civil society. There is ambivalence about the nature of this analogue, and this is important in two, interrelated, ways. First, it means that the contrast between spontaneous self-government and subjection to the power and authority of 'a government' is blurred, and the notion that global civil society is 'self-ordering' is retained. (This converges with the interrelated point, made below, that 'governance' involves self-government.) Second, questions are not begged about the nature of regulatory international institutions and organizations, the functioning of which may be conceived as part of the operation of global civil society.

It is in this context that the significance of the distinction between 'government' and 'governance' should be understood. The essential point is that there can be governance without government. Young (1994: 14) states that 'there is a growing realization that the achievement of governance does not invariably require the creation of material entities or formal organizations of the sort we ordinarily associate with the concept of government'. What then is governance? For Young (1994: 15) it

> involves the establishment and operation of social institutions (in the sense of rules of the game that serve to define social practices, assign roles, and guide interactions among the occupants of these roles) capable of resolving conflicts, facilitating cooperation, or, more generally, alleviating collective-action problems in a world of interdependent actors.

But another important aspect of governance, to be taken up below, is the important sense in which it can be said to involve self-government. This idea is developed, for instance, by Ostrom (1990: 23–8). The basic point concerns the contrast she draws between government imposed from outside and systems of self-government, as ways of dealing with collective action problems (I consider such problems below). On the one hand there is *government*, imposed on those people whose behaviour is to be regulated and which involves coercive external authority; and on the other hand there are rules observed because they are agreed and monitored by those people themselves, which can be described as governance. There are some complex and important issues here,[47] but the paramount notion is that of rules being complied with in the absence of a body with special power and authority to secure that compliance, and of this being due to those rules originating from the people complying with the rules.

This idea of governance involving self-government clearly complements, or further explicates, the idea of civil society being a self-ordering community. And this yields a composite idea according to which democracy can be seen as a system in which government (as normally understood) is absent or has but a subsidiary role, in which there is compliance with rules, but rules which are in some sense self-imposed. Such 'self-imposition' need not involve an actual, definite act of agreement but could be understood as the evolution of rules which gain general acceptance – a process in which general consent to the rules stems from a developing consensus rather than a Lockean social contract. There is also the important matter of monitoring compliance with the rules, and there may be a relatively important role for bodies that in some respects function like governments. But for this to be 'monitoring by the people themselves', such bodies would need to be popularly controlled.

I can return now to my earlier question about the nature of transnational popular control. It can now be shown that there is for the global realm an alternative to the orthodox model of a popular will being implemented through the action of government. Let me consider how this might actually apply. (There is no suggestion here that a global democracy of this kind already exists, or even that it will necessarily come to exist. However, there is the suggestion that this is a model, developed from a consideration of certain existing features of the international-cum-global realm, that is realistic and illuminating and which might well, because of this, come to be actualized. See also note 46.)

The central idea builds on the notion of there being governance in (what is now) the international, or (what will become) the global,[48] realm despite the lack of government: 'it is apparent that governance is by no means lacking in international society, despite the conspicuous absence of a material entity possessing

. . . power and authority' (Young, 1994: 14). Indeed, as Zolo (2000: 83) argues:

> The growing complexity of the international environment tends to produce a systemic situation which James Rosenau (1992) has defined as 'governance without government'. It is a situation in which the absence of a government possessing formal authority (government) co-occurs with a context of extensive phenomena of self-regulative aggregation (governance) of international agents.

For the reasons I have just outlined, this portrayal of the nature of the international and global realm connects up with and further explicates the preceding characterization of global civil society. The idea of governance thus gives content to the notion – explains the operation – of a self-ordering global civil society. Lipschutz, for whom '[r]ules . . . take the place of explicit rule; governance replaces government; informal networks of coordination replace formal structures of command', suggests that '[i]n a sense, global civil society can be seen as part of a growing system of global governance' (Lipschutz, 1996: 252 and 49). But it might well be more accurate to say that in a sense the growing system of global governance is part of global civil society.

It is not always entirely clear just how international organizations and regimes fit into this kind of picture. And here I come back to the ambivalence mentioned above. One of the reasons for the uncertainty and ambivalence is imprecision of meaning. Indeed, it has been said that commentators 'define governance in diverse and often maddeningly imprecise ways' (Bellamy and Jones, 2000: 204). It is suggested here that international organizations and regimes perform three functions. First, following Young, they can themselves be seen as forming a part of the complex of 'the rules of the game'. Perhaps the best interpretation here is that, as suggested above, they should be regarded as monitoring compliance with the rules; and for this to be 'self-monitoring' they should be popularly controlled. But it is the regulatory force of such 'rules' themselves that is primary in understanding governance. I have argued that this comes from their being in some sense self-imposed. To the extent that there is (or there is developing) a global people and community, the 'developing consensus' understanding of this self-imposition, suggested above, could be applicable to the global realm. But now I come to the second function of international organizations and regimes. The 'self-imposition' can also have an element of deliberate choice. The body of rules can be deliberately altered or added to. This can be done by setting up and/or changing the objectives of organizations and regimes; and to the extent that this accords with global popular opinion the new rules can be seen as imposed by the people on themselves.[49] The extent to which the organizations and regimes are democratically structured and responsive to popular opinion is again important here. The third function of these institutions is an aspect of the second. The element of deliberate choice means new policies can be brought in. The operations of the institutions consist in the operation of rules, but these rules execute policy as well as having a 'constitutional' function (see the introductory paragraph to this sub-section). Thus a policy for combating global warming can be operationalized through the appropriate establishment and/or modification of international organizations and regimes. And (to the extent that this obtains) the democratic character of these institutions and the body of rules of which they are a part means such a policy can be democratically chosen and implemented.

But let me return to the 'general acceptance' gained by the body of the rules. This might be interpreted as being induced, or at least maintained, by this assemblage being part of an ascendant ideological position (Lipschutz, 1996: 251). However, this need not be viewed in terms of a Gramscian ideological hegemony so much as an expression of the current situation in a shifting process of contestation or negotiation. A similar interpretation in terms of discourse analysis is developed by Dryzek (1999). Emphasizing the 'sources of governance or order [that are] discursive' (33), he says that '[b]ecause discourses are social as well as personal, they act as sources of order by coordinating the behavior of individuals who subscribe to them' (34). More specifically: 'Discourses are intertwined with institutions; if formal rules constitute institutional hardware, then discourses constitute institutional software. In the international system, the hardware is not well developed, which means the software becomes more important still' (35) He also maintains that because of discursive contestation no one discourse is permanently in the ascendant.

But whether the general acceptance of the rules comes from a process of evolution or of discursive contestation, this can be seen as central to the idea of global self-government.

Transnational and global democracy

I have noted that, to an extent, the recent growth of interest in global democracy as such is a part of the growing interest in the possibilities of transnational democracy in general. This latter derives from reflection on such developments as the growing importance of transboundary issues and processes, the concomitant decline in the importance and power of the state, and the problems these pose for the credibility of orthodox democratic ideas concerning the nature of the people and the possibilities of popular control. Some of the ideas of transnational democracy emerging from this centre around the notion of transnational peoples influencing or controlling international regulatory bodies. Now, it is *global* democracy in which I am interested; and these ideas can be applied to international bodies with a global scope to provide for new mechanisms of popular influence that could become important components of global democracy. Moreover, by helping to weaken the hold of orthodox democratic thought and opening minds to alternative possibilities that are more pertinent in a changing world they help prepare the ground for the development of global democracy. But they do not necessarily have a global reference (indeed, their most prominent test bed has been the European Union[50]); and even when they do they are by themselves insufficient to provide an account of global democracy. As has been shown, such an account must also bring in global civil society and the general processes of governance.

There are two further (interrelated) important points relating to specifically global, rather than generally transnational, democracy. One is the point, already referred to and to be taken up again below, that there are, independently of ideas of global democracy, important notions concerning the (actual or potential) existence of a global *community*. These bolster and reinforce ideas of a global people and global democracy. (There is, of course, mutual reinforcement, and the development of global democracy would also bolster the idea of there being a global community. This is taken up below.) The other point can be seen as an especially significant extension of this; and again it is also referred to elsewhere. This is that the cause of global democracy benefits from the global warming problem itself, since this helps to establish, or to reinforce, the idea of there being a global community – not least (as already remarked) a global 'community of fate'.

To sum up, global democracy is the version of transnational democracy in which I am specifically interested. It is also arguably the most credible version, because of its interconnections with the (to an extent) independently grounded development of global civil society and a global community. As to the form it might take, it is difficult to be specific other than to indicate a picture in which global regulatory bodies would have a democratic constitution and which would operate as part of a global system which could be described as a democratic global civil society incorporating a system of democratic global governance. Such a system would be responsive to global popular demands, through specific popular pressures on the regulatory bodies and the more general functioning of global civil society, manifested most notably in the operations of INGOs.

It is not, in fact, my task to outline in detail the form global democracy might take. Rather it is to indicate its general nature and to go on from this to show the contribution it could make to combating global warming. There are two, interrelated, strands of argument here. First it needs to be shown how the arguments of Chapters 2 and 3 are not invalidated by the fact that it is global rather than national action that is required; that is to say, it needs to be demonstrated that these arguments *can* be operationalized by the development of *global* democracy. Second, attention will be drawn to the additional ways in which global democracy could contribute to the combating of global warming. These mainly concern the ways in which the development of global democracy could help overcome two especially troublesome difficulties in the way of successful action against global warming, namely problems of global collective action and international social justice. It is to these strands of argument that I shall now turn.

Global democracy and global warming

In Chapters 2 and 3 it was argued that the best system of decision-making for dealing with the problem of global warming is one that is democratic. However, there is, of course, a fundamental difficulty with this argument, which at that point I left to one side. The difficulty is that this global problem requires a global response, whereas democratic decision-making is held to exist only *within* states. In other words, since democracy is an intra-state phenomenon it cannot be relevant where the action required is global. If, however, democracy could become a global phenomenon then it would become relevant; and in this chapter I have discussed the nature and possible development of global democracy.

In fact, as already indicated, the existence of global democracy would be doubly beneficial for promoting action against global warming. On the one hand, it would mean that the theoretical benefits of democratic decision-making could in fact be brought to bear on the global warming problem; while on the other hand, there would be a crucial contribution to overcoming major obstacles to the necessary global collective action. Let me look at each of these in turn.

Global application of the benefits of democracy

It can now be seen how the arguments of Chapters 2 and 3 can be made operative. But, more than this, their application to the global situation can involve some supplementation of the arguments and a sharpening of their thrust (and this is apart from the additional benefits of global democracy that are taken up below). Let me consider some essential points.

A fundamental consideration relates to the application of the central anti-guardianship arguments in the full, global, context. The crucial point concerns complexity. There is a greater complexity to the full actuality of the global warming problem than is portrayed in Chapter 2. This might seem to sharpen guardianship claims concerning the need for expertise, and the inabilities of the masses, to cope with complexity. In fact, of course, it gives an added edge to the *anti*-guardianship arguments: to the extent that these are valid in the first place the greater the complexity the greater their force. The additional complexities concern both the phenomenon of global warming and the structure of the situation in which any response is to be generated. The global scale of the phenomenon is an important aspect of its complexity. But this merely reinforces the arguments in Chapter 2 concerning the benefits of democratic decision-making despite – or indeed because of – the nature and role of science in addressing complex natural phenomena. The fact that action must be international rather than national clearly amounts to a massive complication of the situation from which a response must be generated. Here crucial considerations additional to the general anti-guardianship arguments are applicable. These overlap into matters that are taken up below, but in essence the point is that it is still democracy – albeit *global* democracy – that can best deal with the complications. That is to say, democracy here can only be global; and the development of global democracy involves overcoming or removing the obstacles to action inherent in the international realm.

There are also other ways in which there can be a sharpening of the earlier arguments in favour of democracy being best for providing an effective response to global warming. Some of the more important relate to an extension of the argument that the ordinary man is the best judge of his own interests. The basic point here is that the nature of the international situation means that perceptions of interests tend to be structured in terms of national interests. Arguably, however, this applies more to those in government than to the ordinary man, who may be better able to discern that his true interests lie in effective action to combat global warming. There is already some evidence to suggest that public opinion is moving ahead of governments on attitudes towards policies necessary for effective action: according to *The Economist* (7 April 2001: 98), for example, 'public support for action on global warming is strong in both Europe and – say some recent opinion polls – even in America' (for information on some polls relating to Kyoto, see http://www.panda.org/climate/kyoto_games.cfm). And in any case development towards global democracy would enhance this as a decreasing popular focus on narrow national interests would be a concomitant of the transcendence of state-delimited peoples. To push the argument a little further, the existence of a global community in some important sense would be a concomitant of global democracy (see below). And here the argument that the people are the best judges of the communal interest has a particular impact, since combating global warming is in the interests of the global community. Since, à la Rousseau, the communal interest can conflict with particular interests this argument overlaps with that of Chapter 3 to the effect that it is the people who will best promote their long-term interests in combating global warming in the face of competing short-term interests.

Both arguments interlock with the thesis developed in the latter part of Chapter 3 to the effect that a form of democracy is most likely to secure the implementation of really effective global warming policies, by taking proper account of the interests of future generations. And here I can amplify the argument of Chapter 3 about the intergenerational potentialities of democracy. This is partly a matter of the interconnections between the concepts of global and intergenerational democracy: as we saw in Chapter 3 there are conceptual connections between the ideas of an intergenerational and a global community. But the vital point is that since, in a *global* form, democracy really can be brought to bear on the global warming problem, and its capacity to generate the necessary policies that are directed at long-term benefits really can be operationalized, it becomes of real importance that the time span of this capacity be expanded to encompass the interests of future generations. In other words, it becomes of real significance that the developing global democracy could be, and of real importance that it should be, a form of intergenerational democracy.

Other ways in which general arguments are sharpened by focusing specifically on *global* democracy have been touched on previously, but two perhaps should be stressed again here. The first concerns the legitimating and mobilizing functions of democracy. These are in any case especially important in relation to action against global warming due to the need for popular endorsement of policies that may be painful, in the short term at least, and which would otherwise be very difficult to generate and implement. But the obstacles to appropriate policies presented by the nature of the international realm magnify these difficulties,[51] and hence also the need for popular endorsement. The second form of argument sharpening to be stressed again concerns the capacity of democracy to deal with conflicts of values and interests. This characteristic is of especial importance precisely because of the global nature of the policies and the processes of their formation: not only is there the global diversity of cultures to contend with, but the global divisions of interests between North and South raise especial difficulties. I shall take up this latter point below when I consider some issues concerning international social justice.

In various ways, then, the arguments of Chapters 2 and 3 are given added force when the specific nature of *global* democracy is considered. However, there are also other reasons why the development of global democracy would contribute to combating global warming, and which I should now consider.

Additional benefits of global democracy

There are very important ways in which global democracy would facilitate action against global warming, in addition to those that would flow from its specifically democratic aspects. These relate to the fact that the development of global democracy is necessarily interconnected with the development of a global community, and the effect of this on two crucial obstacles to action – collective action problems and problems of international social justice. Let me now briefly consider each of these.

Collective action problems

There is a need for collective global action to respond effectively to the global warming problem. And – as I have several times already had occasion to remark – this means that a fundamental difficulty in the way of generating action against global warming is an aggravated version of the collective action problem. There is, of course, a general problem of collective action (Hardin, 1982); but, notoriously, this is greatly exacerbated by the states system.

The general problem involves a version of the prisoner's dilemma,[52] and it is that in many circumstances, where there is no effective coordinating mechanism, it is in the interests of agents individually to do that which is against their collective interests. In the international realm this problem is magnified by the fact that the agents are powerful states, the whole logic of their situation is such that they are focused overwhelmingly on their individual (national) interests and there is in any case often little or no perception that there is a competing substantive collective interest.[53] This basic difficulty is helpfully surveyed by Hurrell and Kingsbury (1992a). Their quote (6–7) from Falk (1971: 37–8), writing at an earlier period of environmental concern, brings it into sharp relief:

> A world of sovereign states is unable to cope with endangered-planet problems. Each government is mainly concerned with the pursuit of national goals. These goals are defined in relation to economic growth, political stability, and international prestige. The political logic of nationalism generates a system of international relations that is dominated by conflict and competition. Such a system exhibits only a modest capacity for international co-operation and co-ordination. The distribution of power and authority, as well as the organization of human effort, is overwhelmingly guided by the selfish drives of nations.

Of course, a central feature of the situation so described is that the international system has no overall government. Government – and government *enforcement* of what is in the collective interest, so that this does become what is in the individual interests of agents – is the classic solution to collective action problems. Dunn (1994a: 13) conveniently summarizes the basic points when he says; 'The problem of collective action permeates all politics' but is 'peculiarly intractable where there is little realistic prospect of creating an effective enforcement agency and where the rational appeal of seeking to free ride[54] is often devastatingly apparent.'

This, then, is the 'international collective action problem' – the intractable form of the general collective action problem in the international realm, where there is no enforcement agency to secure the collectively rational outcome. And – in addition to, and exacerbating the other problems – it is widely believed that this is what stands in the way of generating action to combat global warming. The problem is aggravated by the fact that, while it is seen that there is indeed a genuine collective interest in combating global warming, the competing individual, national, interests, which are largely economic, can seem more urgent and more important:[55] this is certainly how President Bush sees it. Although there is a parallel and overlap here with the issue of short- versus long-term interests discussed in Chapter 3, there is also a difference. The difference is, precisely, that the long-term interest now relates to benefits that suffer a double bind, because not only are they in the (perhaps somewhat distant) future but, due to of the collective action problem, also because it is irrational for individual agents to attempt their realization.

This problem has bedevilled attempts to tackle the global warming problem from Rio through to Kyoto and beyond. It is true that the growing realization of the existence and seriousness of the global warming issue has modified the problem by greatly strengthening the claims of the collective interest. And the progress from the non-binding UNCED agreements to the binding Kyoto Protocol might be taken to show that the problem is being resolved or transcended. This, indeed, would be the view of 'neo-liberal institutionalists'. Against this, however, it can be maintained that the Kyoto Protocol – especially as amended at Bonn in July 2001– is too weak; and that anyway the stance of the United States undermines the whole Kyoto process. And the fundamental reason for this, many would contend, is the collective action problem as indicated above. To put the point another way, the contention is that even if (as in the premise of this book) it is granted that collective global action to combat global warming is in the best interests of everyone, the collective action problem will continue to prevent such action taking place.

The theoretical reasoning underpinning this contention can, however, be questioned. That is to say, it can be questioned whether there really is an international collective action problem. The idea that there is, rests essentially on the Hobbesian analysis[56] that it is only the presence of a government that can solve the collective action problem, and that this is what is necessarily lacking in the international realm. But there is a recent literature questioning the reality of the 'tragedy of the commons' version of the collective action problem,[57] which suggests that the problem can be solved without bringing in government. Ostrom's (1990) book is very important here, and the line of argument has been applied to the international realm (Oye, 1986). However, as Young (1994: 16) puts it, such solutions do still involve *governance*. We are left, then, with a situation in which effective action against global warming will require global governance; or, at least, a system of global governance in matters which relate to combating global warming. Now, I have just rejected the adequacy of what optimistic neo-liberal institutionalists would see as (at least the beginnings of) such a system, viz the global climate regime being established by the Kyoto process. However, it can now be argued that this inadequacy is a manifestation not of an insoluble international collective action problem but of the lack of sufficient governance. That is to say, the problem is no longer the inevitable lack of government in the international realm but the lack of sufficient governance, the point being that while the former may be insoluble the latter is not. It remains true that the problem has not yet been solved. And, arguably, to be successful a global climate regime must be a part of an overall system of global governance. Only in this way can there be sufficient 'rules of the game', with sufficient potency, to control adverse forces. To put the point another way, only a full system of global governance can modify the international realm sufficiently to overcome its inherent collective action problem.

This is where I come back to global democracy, since, as I have already argued, that would provide just such a system of global governance. Indeed, given that, according to my earlier analysis, a system of global governance (as distinct from government) will necessarily have democratic features;[58] and given the other factors I have already considered favouring its development, it is arguable that only global democracy will bring such a system. In other words, to be successful a climate regime would need to be developed as part of a system of global democracy. (In fact, according to the argument of this book a global democracy may well be centred on a democratic climate regime; but then the success of the regime would stem from its *being* a system of global democracy.)[59]

This is yet another reason, then, for arguing that the development of global democracy is the best hope for procuring sufficient action against global warming.

Problems of international social justice

I have focused above on the governance involved in a system of global democracy as the reason why such a system would resolve problems of collective action. But another reason is that the development of global democracy would involve the development of a global community. (This in fact also partly explains the existence of governance itself.) There are various point here, on some of which I have already remarked, but there is one in particular that should now be mentioned. This concerns the problems caused by the interlinking of issues of international social justice with the other difficulties relating to generating action against global warming, and how these might be resolved by the development of a global community. I have developed this argument elsewhere (Holden, 2000b), so it will only be briefly outlined here.

The premise is that attempts to combat global warming are made more difficult because of the important considerations of social justice that they raise, and that these are complicated and made more intractable because they are issues of *international* social justice. The argument is that some kind of global community would be a crucial help in surmounting these difficulties and that this would be a concomitant of the development of global democracy. I shall glance first at some of the inherent problems of international social justice, before indicating the nature and importance of the issues of international social justice involved in the global warming problem and how the resolution of these issues depends on overcoming the inherent problems. I shall then look at how some kind of global community would help in overcoming the problems, and then come back to the connections between the existence of such a community and the development of global democracy.

The inherent problems with international social justice concern the extent to which ideas of social justice may be inapplicable to the international realm. The central point is the traditionally tight linkage between such ideas and that of a state community. The central notion – further developed in Holden (2000b: 197–8) – is that only membership of such a community makes feasible the modification of self-interested behaviour that is involved in socially just actions. The basic ideas here are that this modification stems from the special moral relationships and ties of identification and solidarity among members of a national community. There is, too, an overlap with the collective action problem, since these ties are powerfully reinforced, or made operative, by the fact that in a political community this problem is solved so that in a whole range of cases selfish behaviour is not the only rational behaviour it would otherwise have been. In the *inter*-state system of the international realm, on the other hand, there is no community and therefore no room for social justice. It is true that to an extent corrective justice is applicable in the states system, but this is not the same as social justice (Holden, 2000b: 199). 'International' social justice is a chimera and demands that social justice should be globally applicable are really demands for *global* social justice in a global community.

Let me now turn to the question of the issues of international social justice that arise in relation to the problem of global warming. And we should notice first how and why these are so important. It is not only that resolving them is morally desirable, but also that this is a necessary condition for obtaining action against global warming. Only policies accepted as socially just will secure the general agreement necessary for their implementation. As Paterson (1996: 182) points out, the 'general debate on global warming, as well as that on other environmental problems, is replete with assertions that agreements need to be equitable in order to produce effective responses to those problems'. However, the difficulties with the idea of *international* social justice frustrate efforts to address these issues properly; or, to put the point another way, the prioritizing of state sovereignty and the national interest mean there is a denial that there are genuine issues of social justice involved. It is true that issues of corrective justice also arise: it is commonly held that because some rich Northern states have been primarily responsible for global warming, justice – corrective justice – demands that those states bear the costs of preventing, or mitigating, the harm it will do to other states. And there is some confusion, and a blurring of the distinction, between corrective justice – which *is* clearly applicable to the international realm – and social justice. For this, and other reasons, lip service is paid to the idea that there are genuine issues of social justice involved. But in fact, when it comes to the crunch, there is little real acceptance of their validity and there is a reluctance to take them seriously (Holden, 2000b: 200–1).

In order to better understand the problems here, and the need for their resolution by the development of a global community, we should see what the relevant issues of social justice are. (Noting that they can only be fully understood against the background of, and as linked into, larger issues of the justice – or, rather, the injustice – of the distribution of wealth and power between the North and the South (Shue, 1992).) These issues primarily concern the distribution of the economic costs of combating global warming. More specifically, the questions involved are those such as which countries should take what (economically harmful) abatement measures, and who should pay what for developing new environmentally friendly energy technologies. There are arguments from social justice to the effect that because it would be unfair to expect the Southern countries to forgo the economic development that has made the North rich, and because the Northern countries have the necessary economic resources and expertise to help, the North should transfer resources to the South to enable them to develop economically in an environmentally friendly way. This is the basic international social justice argument relating to the global warming problem, though it does raise, or link in with, other issues (Holden, 2000b; Paterson, 1996). One of these is that social justice also involves a consideration of the interests and claims of future generations, and I shall take this point up below.

As I indicated just now, partly because there is an overlap or confusion between this basic argument and an important corrective justice argument, it has had some effect. Possibly this 'dual grounding' gives it some extra force. And, indeed, as we have seen, some provision for transfers was written into the Rio Climate Convention and has been developed since. However, as subsequent difficulties – most recently with the Kyoto process – have shown the argument has not been properly effective. This surely indicates that under pressure the grounds come apart and issues of international social justice become excluded as illegitimate. Hence they fail to bolster – let alone press beyond – the claims of corrective justice against the assertion of what are deemed vital national interests. It is accepted here that justice, both social and corrective, does demand greater help for Southern countries by Northern countries; but at this juncture the central point is that it is a necessary condition for reaching agreement on action against global warming that the claims of social justice should be satisfactorily addressed. So we now need to look at how this might be achieved – how global democracy might bring about a validation of considerations of 'international' social justice by translating them into issues of 'global' social justice.

I focused above on the extent to which problems concerning international social justice arise from the international realm being constituted of separate communities rather than itself being a community. The development of global democracy would help resolve these problems to the extent that it would convert the international realm into a global community, in relation to which issues of what would then be global social justice would clearly be applicable. I have remarked many times on the connections between the development of global democracy and the establishment of a global community, but four points should be made here. First, while it is clear that these connections involve a relationship of reciprocal causation, the development of global democracy *is* one half of this relationship and it remains the case that it is necessary for the development of a global community. Second, the development of a global community would involve the undermining of that overarching identification with national communities and the prioritizing of narrow national interests, which is a concomitant of the states system, and its (at least partial) replacement, on a global scale, by those feelings of communal identity and solidarity that make socially just actions feasible. Third, the global community developed as a concomitant of global democracy would be a political community, so the feasibility of socially just actions would be reinforced by the avoidance of collective action problems that global democracy's system of global governance would bring. Fourth, it should again be remembered that, besides calling for the development of a global community and global democracy, the need to cope with the problem of global warming does itself encourage such developments.

The development of global democracy, then, may well be a

precondition for the satisfactory resolution of the problems of what would otherwise remain intractable problems of international social justice. But it is not only 'geographical' problems of social justice that are raised by the global warming problem. As we saw in the last chapter, important issues of intergenerational justice are also involved. Now, it is true that, unlike the 'geographical' issues, the latter are not engendering problems that are overtly intruding on, and impeding, efforts to reach agreement on action to combat global warming. Nonetheless, they do directly affect the possibilities of arriving at really effective action. As we have seen, just treatment of future generations would be a necessary feature of such action; but, in the absence of countervailing factors this is action that may well be prevented, since it is unlikely that a focus on intergenerational justice will surmount the 'self-interest' of the present generation. It is here that a global democracy that is also an intergenerational democracy would be invaluable. As I have already argued, intergenerational democracy is the best means for securing that justice for future generations that, besides being a basic moral imperative, is a sine qua non of really effective measures to combat global warming.

I have now considered most of the major arguments for maintaining that democracy, and more specifically global democracy, is the best means of promoting action against global warming. It is time now to try and pull some of the threads together and briefly to suggest some tentative conclusions regarding the extent to which it is sensible to treat the development of global democracy as a realistic possibility.

Notes

1. The meaning of the term 'international' includes the idea of a system of separate states: 'Use of the term "global" implies an ideological and methodological orientation based on conscious rejection of the term international. . . . This involves an attempt to move away from the traditional, realist, approach to international politics, with its focus upon interstate relations . . . in order to encompass both a more inclusive agenda and a range of non-state actors' (Bretherton, 1996: 2).
2. It might be argued that the concept of 'rule' is logically tied to that of 'the state'. However, there is some (albeit hazy and uncertain) overlap between

the concepts of 'rule' and 'governance'. The question of the relationship between 'government' and 'governance' is taken up later.

3. Of course, there have not always been states to *be* instruments of rule: the tight connection between states and ruling is a characteristic of the modern states system. Previous to its emergence ruling was carried out by – and diffused among – a diversity of persons and institutions. (Some in fact see the kind of contemporary developments on which I am now focusing as perhaps leading to a situation in some ways analogous to this 'pre-Westphalian' system (Held, 2000: 27).) Nonetheless, democracy has up until now been tied to the state, or a state-like entity: modern democracy only emerged after the consolidation of the modern states system and the only other manifestation of democracy occurred in ancient Greece where the *polis* system was in certain key respects similar to the modern states system.
4. In contradistinction to this, and interestingly in the present context, it has been argued that there *is* a logic within the framework of democratic theory which does provide a solution, and one which lends support to the idea of global democracy: 'as Archbishop Temple pointed out, "The abstract logic of democracy may tend towards cosmopolitanism" for "the people" have no theoretical interest in the boundaries of states which dynastic jealousies had caused to be so bloodily defended and extended over the centuries' (Heater, 1990: 56). This sort of idea has heretofore received little serious attention within democratic theory.
5. It might be argued that it is not the concept that is under threat of dissolution so much as existing alleged 'nation states' for failing to embody the concept (Hont, 1994). Even so, the perception of the *existing* international realm as essentially divided into nation states would still be undermined.
6. For example by Spitz (1979), Whelan (1983) and Kimber (1989). See also, Saward (1998).
7. There are, indeed, important issues of democratic control relating to foreign policy. There is, of course, a much-discussed issue concerning the possible and desirable extent of influence over, or control of, a state's foreign policy by the electorate of that state. But my interest here would be in questions concerning influence on a state's foreign policy by those outside the state who are affected by it. This raises a fundamental issue since it would appear absurd to assert, for example, that where state A is in a situation of conflict (perhaps even actually or imminently at war) with state B, the people in state B should have a right to influence or control the actions of the government of state A. This is, indeed, one of the criticisms of May's definition; and it might be maintained that this is precisely where the logic of the argument for the all-affected principle leads. However, this is to ignore the crucial point concerning the role of the all-affected principle in *transcending* the states system. The logic of the

all-affected conception of the people points to a reduction in the importance, and perhaps the eventual dissolution, of states; and where states are less important or non-existent so too is hostility between them. To put essentially the same point another way, the logic of the all-affected conception of the people points towards reducing or dissolving the distinction between governments' domestic and foreign policies.

8. Another reason, of course, is the tight conceptual relationships between the established conceptions of 'the government', 'the state' and 'the people' such that it is *logically* true that a government of a state can have certain kinds of duties only to the people of - 'the people' as delimited by – that state. But, again, this is to miss the point that it is precisely established conceptions that are being critically examined (see also the previous note and the main text below).
9. I am interested here in cases where the all-affected principle *extends* the range and number of relevant persons beyond those included in the traditional people. These are the most important cases. But it should perhaps be noted that on occasion there can be cases where the principle *limits* those who are relevant to only some among the traditional people: and this might be seen as a defect of the principle.
10. Within orthodox democratic theories there are varying accounts of the relationship between the government and the people, but they can all be said to involve the people 'controlling' the government, at least in the widest sense that is implied by the definition of democracy as a 'system in which the people, positively or negatively, make and are entitled to make, the basic determining decisions on important matters of public policy' (Holden, 1993: 8).
11. Again, for the sake of simplicity the reference is only to the spatial parameters. As we have seen, although the notion of intergenerational democracy involves different temporal sets of persons there is a sense in which they constitute an intergenerational community. Moreover, such an idea links with the notion of a global spatial dimension, so we have here a notion of a people which is spatially fixed but which also has a temporal dimension.
12. There is now a large literature on globalization. For an introduction to the topic – and for contending views on the nature and significance of globalization and its implications for global democracy – see Holden (2000a). Recent works that might be mentioned here include Held *et al.* (1999), Hirst and Thompson (1999) and Mickelthwait and Wooldridge (2000). Higgot and Payne (2000) contains a comprehensive collection of articles (1991–2000).
13. This illustrates again that while global warming is a problem, not least because it requires a global response, the very existence of this problem does itself also help generate the conditions for such a response.
14. It is accepted here that such diminution is taking place. There is, however,

ongoing controversy regarding the extent to which state autonomy is or is not being diminished by globalization: see, for example, Hirst and Thompson (2000), Perraton (2000) and Golub (2000).

15. This applies increasingly to domestic matters as well as events and processes in the international realm. Indeed, an important aspect of globalization here is the way in which this distinction is breaking down and the extent to which 'domestic' (especially economic) matters are becoming meshed into (sometimes global) processes extending beyond a state's borders in a way which puts them beyond proper control by that state.
16. This illustrates again the way in which global warming at the same time as it poses a problem makes some contribution to its solution. Here it is shown that by being a *global* problem global warming contributes generally to a global consciousness, and in particular to an awareness of the need for *global* action. Paradoxically perhaps, this is reinforced by a perception of the lack of capacity of states for appropriate action (see the next sentence in the main text). In sum, global warming forms an aspect of that very globalization that may – not least by encouraging the development of global democracy – facilitate action to combat it.
17. I shall also comment on the efficacy of existing attempts to overcome these difficulties.
18. Globalization may, though, provoke efforts to obtain more efficacious collective action. This is, indeed, one of the themes in the global democracy argument below.
19. There is, of course, a fourth possibility: collective action by states to obtain what none (or only some) of the separate peoples want. But this is of no relevance in a discussion of the possibilities of *democratic* action. It is noteworthy that according to many of its critics it is this (among other things) that makes the European Union undemocratic.
20. Where (a properly functioning) nation and/or state could supply or inspire the communal identity.
21. A distinction is sometimes drawn between 'organizations' and 'institutions': 'Whereas institutions are sets of rules of the game or codes of conduct defining social practices, organizations are material entities possessing offices, personnel, budgets, equipment, and, more often than not, legal personality. Institutions affect behavior of these actors' (Young, 1994: 3–4). However, this distinction is not always clear cut, and both can be instruments of governance ('Institutions affect the behavior of these actors'). The general term 'bodies' is here used to cover both. In what follows there will not be too pedantic an attachment to this distinction, but use will be made of 'bodies', 'institutions' and 'organizations' as seems appropriate.
22. And, of course, the remaining power of states and the states system is the other key reason for the necessity of such regulation. The impotence of

states in the regulation of transboundary processes is as much to do with their being entrapped in a system in which an ingrained prisoner's dilemma situation means still relatively powerful states pursue short-term selfish interests – and are checked by the countervailing power of other states – as it is to do with a decrease in the intrinsic power of states. (The prisoner's dilemma situation is discussed in the main text below.)

23. The importance of international regimes as regulatory mechanisms has, of course, been increasingly stressed by those who believe that regulation by international organizations is possible. See, for example, Krasner (1983) and Rittberger and Mayer (1993). For a survey relating especially to environmental issues see List and Rittberger (1992).
24. Use of the term 'international' implies a realm that is fundamentally constituted by states and their interaction. To the extent that transnational democracy developed, states would become increasingly less important. A similar point is often made by suggesting that the era of globalization is tending towards the creation of a 'global' rather than an 'international' realm.
25. An example is provided by the way in which the International Sea-Bed Authority manages mineral resources on the floor of the deep ocean – resources which are conceived to be 'the common heritage of mankind' – for the benefit of the community of nations as a whole (see, for example, Brown (1988: 270–1). True, the direct reference is still to nations rather than people – to the community of nations rather than the community of mankind. However, at least there is reference to accountability to a body beyond the states of its governing assembly. Moreover, there is indirect reference to the community of mankind since the idea of 'the common heritage of mankind' is invoked. Significantly in the present context, in the subsequent edition of his book, Brown widens the argument to include management of 'international commons' generally and 'the new initiatives in international accountability launched' at the Rio 'Earth Summit' in particular (Brown 1995: 264). See also the main text below.
26. See also the analysis in (Brown 1995: 255-6), including remarks on 'the political and legal arrangements for implementing the basic accountability principle'.
27. See also Saward's critical evaluation of Held's ideas: Saward (2000).
28. Concern over rioting at recent meetings of world leaders (which itself can be seen as indicative of a 'democratic deficit' – see main text below), with the death of a demonstrator at the G8 meeting in Genoa in July 2001, has cast doubt over the feasibility of future meetings. However, there is likely to be a change in the type of venue rather than the demise of the institution.
29. There could, of course, be reforms that *replaced* the existing body with one which is more representative: this is the line taken by the Commission on Global Governance (1995: 155–60).

30. Equal representation of states is at present frustrated by the more salient role for the permanent members of the Security Council. But (apart from the points made in the main text) that equal representation could achieve a United Nations that was only 'democratic' in a bogus sense is illustrated by the fact that among the states equally represented would be those that are *un*democratic! However, the United Nations could become, or the General Assembly be supplemented by, an 'international democratic assembly' of all – and only – democratic states. This is advocated by Held (1995: 273–4), though he moves from the existing notion of an assembly of delegates appointed by states' governments to the idea, mentioned in the main text below, of an assembly of delegates directly elected by, and accountable to, the peoples of the states.
31. Use of the term 'delegate' is not meant to beg questions about whether those elected should be only the mouthpiece of those who elected them, or whether they should have a significant degree of discretion to use their own judgement (see, for example, Holden, 1993: 73–7). Rather, 'delegate' is used because this is the term frequently used to refer to states' representatives at international organizations, conferences, etc.
32. Also, in the process leading up to the Earth Summit (from the Stockholm Conference of 1972 onwards) the leading actor was a UN agency: the UN Environment Programme (UNEP). (In the case of the Earth Summit, 'as befitted the political importance of the subject, the negotiation took place under the aegis of the General Assembly itself' (Brenton, 1994: 185). More specifically, this was 'under the auspices of . . . the Intergovernmental Negotiating Committee on Climate Change established by the United Nations General Assembly specifically to develop a climate regime' (Young, 1994: 166).)
33. An issue which cross cuts, and overlaps with, questions of transnational democracy is that of international social justice and the relationship between the (rich) North and the (poor) South. This will be taken up below. In the case of the CSD, too, representatives of states 'are elected as members with due regard to equitable geographical distribution' (Agenda 21, quoted from *Earth Summit '92*: 231).
34. The UN Conference on Environment and Development (UNCED) is the official title of the 1992 Rio Earth Summit. Agenda 21, one of the UNCED agreements, 'is intended as the action plan for achieving sustainable development' (Grubb *et al.*, 1993: 17).
35. This would now be regarded as a narrow and technical sense of the term. Today, terms such as 'pluralist democratic theory' would be normally be taken as referring to general ideas concerning democracy and diverse values and cultures; see, for example, Bellamy (1999).
36. On the distinction between interest and cause groups see, for example, Baggot (2000).
37. This is true even of *global* democracy – where there might seem to be a

parallel with traditional democracy, in that there is a pre-determined terrain to which the concept of democracy is to be applied. This, however, would be to ignore the point that essential questions are precisely *whether* the concept might be applied to global processes; and, if so, what the relevant processes are.

38. Moreover, these may be communities that are especially conducive to democratic activity. As Dobson notes, 'Dryzek makes much of the "new social movements" which he believes represent "real world approximations" to the ideal of discursive democratic practice' (Dobson, 1996: 147, quoting Dryzek, 1990: 49).
39. According to the now dominant conceptions there is a distinction between civil society and the market. (And, in view of the noted overlaps between 'globalization-from-below', global democracy and global civil society, this ties in with our contrast between all three and global capitalism.) But it is not just some Marxists who would see things otherwise. Barry (1999: 237) points to 'two senses of civil society'. The first is 'what one may call "the traditional liberal" understanding of civil society which identifies it in the first place with . . . market society'. In the other, or 'post-liberal', sense the term refers 'to associations that are independent of the state and the market economy'. But, he adds, it 'is this second understanding of civil society that is closest to green concerns'.
40. The UN's view of what other elements are to be included is outlined by Segall (1997: 339), who indicates the particular meaning of 'civil society' at the UN: 'There it encompasses all NGOs; all religious, academic, business, professional, media, trade union, local government, and indigenous organizational bodies; and other groups and networks distinct from the state's legislative and administrative power.'
41. But see note 39 above.
42. It can be seen as a necessary condition both by providing a satisfactory account of 'a people', and, à la Tocqueville, by providing the kind of democratic society and culture necessary for democracy's functioning and maintenance: 'A democratic state is widely held to be inconceivable without a democratic society' (Hirst, 1996: 97); see also, for example, Cohen and Arato (1992: 19 and 117). This notion can shade into the idea that democratic civil society *is* a democracy (see main text below) rather than one of its necessary conditions; and demands for further democratization of civil society may be associated with both.
43. Such as those associated with anarchism and Marxism. Burnheim (1985) also has an interesting non-governmental account of 'demarchy' (though whether 'demarchy' is a form of democracy is arguable: see Holden (1988)).
44. 'Theorizing of' can here refer to two (different but interconnected) propositions, and both are being affirmed. On the one hand there is a proposition about giving an account of a diversified community and on the

other there is a proposition about such a community's existence. The first is a necessary – though not a sufficient – condition of the second. Similar considerations apply in similar contexts.

45. There is a further, and cross-cutting, consideration that I shall not go into here. This is that discussions of global democracy do not necessarily centralize the notion of the exercise of power by a global people. Thus Held's discussions of global, or cosmopolitan, democracy – for example, Held (1995, 2000) – focus more on the global rule of law and relationships between a multiplicity of transnational organs of governance and relevant transnational groups of people, than on the idea of action by a single global people.
46. There is no firm implication here about the salience of global civil society, though there is at least the contention that a nascent global civil society exists and that there is every likelihood that it will become more salient. The implication could be that the global realm already is, or is in the process of becoming, essentially a global civil society; or it could be that global civil society is a phenomenon within, and a challenge to, a realm dominated by states and/or global capitalism. This latter notion ties in with Falk's idea of 'globalization-from-below'.
47. Two important points might be noted here. One is that we are here on the cusp of a division between two lines of thought. On the one hand, there is a tie-up with non-governmental understandings of democracy (to which I have already referred); but on the other hand, there is an affinity with Rousseauan ideas of *government*, albeit the self-imposed government of rule by a general will. There are resonances of both in the ideas of governance as involving self-government that are discussed in the main text. The other point is that there may well be an oversimplification, and blurring of issues, in the notion of the 'external' imposition of government. Ostrom is primarily talking about small communities or enterprises in relation to which national government is 'external'. However, in relation to the identity of members of the community or enterprise as national citizens, the national government is no longer external in the same way. (It may be, though, that there is likely to be a more straightforward application of the notion of externality in the international realm.)
48. There is quite an important issue, that I have not the space to go into here, concerning the change in 'objects' of governance associated with a change from international to global. In the former it is typically (but no longer solely) states that are to be regulated whereas in the latter it is non-state actors – including individuals.
49. This is different from a whole body of rules being a deliberate construction – where they can have regulatory force only as legislation (and, in a democracy, with the added legitimacy of being popularly endorsed). In the global arena, of course, I am assuming the absence of legislation as such, and in the present case the regulatory force comes from the extant body of

rules into which the new ones are being absorbed. Popular endorsement ensures this absorption and augments the differently based consensus characterizing the whole body of the rules. Of course, this distinction can become very hazy, which reflects the haziness of the distinction between government and governance.

50. Although it must be said that the ideas and structures of European Union have often been taken as possible models for structures of global integration; and the issue of their democratization has similarly been taken as providing illuminating parallels for considerations of possible global democratization (for example, Golub, 2000: 180).
51. There are really two different considerations here. On the one hand, there are the familiar obstacles to concentration on a global common good presented by the nature of the existing international realm – which the development of global democracy would overcome – and on the other hand, there are the obstacles to the realization of effective policies that would still be presented by the scale and complexity of even a truly global realm.
52. 'A prisoner's dilemma is a situation in which the rational choice for each individual in a set prevents the set of individuals from realising joint gains from co-operation' (Weale, 1992: 41). The literature on the prisoner's dilemma and the logic of collective choice is extensive. See, for example, Mueller (1979), Hardin (1982), Taylor (1987).
53. The point here is that a clear perception of a superior collective interest is probably a necessary condition for – and certainly an important factor in – any amelioration of a collective action problem. Of course, as I have remarked many times, a notable feature of the global warming problem is in fact that exceptionally, it is, giving rise to a clearly perceived collective interest that, at least arguably, is superior to competing individual state interests.
54. The extent to which there is an opportunity to free ride in respect of global warming policies is variable. On the one hand, there is the opportunity for small states, which emit relatively small quantities of carbon dioxide, to gain the benefit of global reductions by leaving it to other states to cut emissions. On the other hand, states which emit large quantities could not really benefit by leaving it to others, since that state's continued emissions would prevent there being a sufficient reduction of the global total. President Bush is not so much gaining a free ride as preventing all riding, since 25 per cent of the world's total carbon dioxide emissions come from the United States.
55. There is a double aspect to this. Not only may the absolute economic costs of cutting carbon emissions undermine the attraction of the competing collective interest, but the relative economic costs of cutting when other countries do not, strengthens the need to promote individual, national, interests at the expense of the collective interest. It should also be noticed that the structuring of perceptions along the lines of national interests

itself tends to obscure perceptions of a collective interest (a difficulty which, as I have previously remarked, would be overcome by the growth of global democracy and community).

56. There is a 'Hobbesian analysis' of both the collective action problem among individuals and among states in the international realm. Although he is primarily concerned with the former, an explicit formulation of the latter, which is essentially parasitic upon the former, is also to be found in Hobbes, and it is central to the realist tradition of international thought.
57. The idea of 'the tragedy of the commons' has been very widely used in discussions of collective action problems since the publication of Hardin's (1968) influential article of that title. It centres on the predicament of herders who graze their animals on pasture which is open to all, but in doing so are trapped in a situation where their individually rational behaviour results in collective ruin as the pasture becomes overgrazed. For a useful summary, see Ostrom (1990: 2–3). As mentioned in the main text this book also contains a critical analysis.
58. The essential point is that, according to my earlier analysis, a system of governance is in an important sense a system of self-regulation (although to be fully a democracy such a system must also provide for a more detailed popular input into specific policy-making aspects of the functioning of the institutions of governance).
59. Although many matters would be, of course, beyond its purview, a democratic global climate regime could nonetheless in itself constitute a global democracy since there would be a fairly comprehensive range of matters that it would need to regulate. After all, in orthodox democratic theory there are many matters that are meant to be beyond the purview of a *liberal* democratic state.

Conclusions

It has been argued that global democracy offers the best hope for dealing with the problem of global warming. But how realistic is such an argument? Many will reject talk of the development of global democracy as utopian fantasizing and dismiss reliance on such a development for generating action to combat global warming as a naive delusion. How valid is such criticism? Space does not permit a detailed exploration of the issues raised, so I shall simply indicate two important lines of counter-argument.

The first kind of argument rejects the charge of utopianism, but accepts that theorizing about global democracy is 'ideal theory'. However, it insists that there is a valid and valuable role for such theory. Beitz (1979: 156) – in a different, but in some respects comparable, context – puts the point well:

> Ideal theory prescribes standards that serve as goals of political change in the nonideal world. . . . The ideal cannot be undermined simply by pointing out that it cannot be achieved at present. One needs to distinguish two classes of reasons for which it may be impossible to implement an ideal. One class includes impediments to change that are themselves capable of modification over time; the other includes impediments that are unalterable and unavoidable. Only in the second case can one appeal to the claim of impossibility in arguing against an ideal. . . . [Ideal theory] requires only that the necessary

> changes be possible, and it is at least not demonstrably false that this is the case.

To this it can be added that efforts in the 'nonideal world' to deal with the problem of global warming have not been notably successful and that if it is to be dealt with at all then new ideas and practices are required.

This line of argument, however, might concede too much. And this brings me to the second line, which maintains that to theorize about global democracy is not necessarily to engage in 'ideal theory'. The burden of much of the argument in Chapter 4 was that there are developments in the 'real world' that are undermining the existing international order and which point towards the possible development of global democracy. And there is influential support for this viewpoint. Chris Brown (1998: 113), for instance, refers to 'David Held's work on . . . new democratic forms which would operate at a global level', and adds that '[i]t may be that the time has come for the issue of global institutional reform to shake off the stigma of utopianism. . . .' And Andrew Linklater, referring specifically to Europe but with an eye to more general conclusions, argues that 'complex processes of social change reveal that the notion of cosmopolitan democracy is no longer fanciful' (Linklater, 1998: 121). There is another dimension which must be added to this line of argument. This focuses on the exceptional 'fluidity' of the post-cold war world and the way the unpredictable and 'unrealistic' developments have already occurred. Richard Falk (1999: 124), for example, talks of an 'aspirational future [that] seems like a pipe dream at present'; but, he goes on:

> only a decade ago so did a post-cold war world and a post-apartheid South Africa. How could such developments have occurred without the presence of concealed, yet latent and formidable, social forces committed to images of drastic reform that were condescendingly dismissed in realist circles as 'utopian and irrelevant?'

Perhaps this general line of thought is best summed up by Anthony McGrew (1997a: 241):

> Although normative theory is concerned primarily with explicating and analysing what is desirable, the principles underlying what ought to be or should be the case, it is neither necessarily wildly utopian in nature nor divorced from an

> understanding of contemporary historical circumstances. On the contrary it derives its intellectual credibility from an understanding of both '. . . where we are – the existing pattern of political relations and processes – and from an analysis what might be: desirable political forms and principles' (Held, 1995: 286). To discount normative theory simply on the grounds that it trades in ideas or projects which, under existing historical conditions, may appear politically infeasible is to accept a deterministic view of history. But '1989 and all that', completely unanticipated as it was, is a solemn reminder that prevailing assumptions about what appears politically feasible are often a feeble guide to history's possibilities.

And now this viewpoint is perhaps given added impetus by considering the fallout from the dreadful events of 11 September 2001. At the time of writing it is unclear what will be the eventual impact on the international order of the attack on the World Trade Center, but it is generally agreed that 'everything has changed'. This illustrates again the possibilities of fundamental change in the contemporary world.

In conclusion, then, the possibilities of the development of global democracy, and hence of it making an invaluable contribution to dealing with the problem of global warming, should not be discounted.

Bibliography

Achterberg, W. (1996), 'Sustainability and associative democracy', in W. M. Lafferty and J. Meadowcroft (eds) (1996b), *Democracy and the Environment*. Cheltenham: Edward Elgar.

Agius, E. (1998), 'Obligations of justice towards future generations: a revolution in social and legal thought', in E. Agius and S. Busuttil (1998), *Future Generations and International Law*. London: Earthscan.

Agius, E. and Busuttil, S. (eds) (1998), *Future Generations and International Law*. London: Earthscan.

Archibugi, D. (1995), 'From the United Nations to cosmopolitan democracy', in D. Archibugi and D. Held (eds) (1995), *Cosmopolitan Democracy*. Cambridge: Polity Press.

Archibugi, D., Balduini, S. and Donati, M. (2000), 'The United Nations as an agency of global democracy', in B. Holden (ed.) (2000a), *Global Democracy: Key Debates*. London and New York: Routledge.

Archibugi, D. and Held, D. (eds) (1995), *Cosmopolitan Democracy*. Cambridge: Polity Press.

Archibugi, D., Held, D. and Köhler, M. (1998a), 'Introduction', in D. Archibugi, D. Held and M. Köhler (1998b), *Re-imagining Political Community*. Cambridge: Polity Press.

Archibugi, D., Held, D. and Köhler, M. (eds) (1998b), *Re-imagining Political Community*. Cambridge: Polity Press.

Arrow, K. J., Cline, W. R., Maler, K.-G., Munasinghe, M., Squitieri, R. and Stiglitz, J. E. (1996), 'Intertemporal equity, discounting, and economic efficiency', in J. P. Bruce, H. Lee and E. F. Haites (eds), *Climate Change 1995: Economic and Social Dimensions of Climate Change* (contribution of Working Group III to the second assessment report of the Intergovernmental Panel on Climate Change), Cambridge: Cambridge University Press.

Axtmann, R. (1996), *Liberal Democracy into the Twenty-first Century*. Manchester and New York: Manchester University Press.

Baggot, R. (2000), *Pressure Groups and the Policy Process*. Sheffield: Sheffield Hallam University Press.

Barnaby, F. (ed.) (1991), *Building A More Democratic United Nations*. London: Frank Cass.

Barry, J. (1994), 'The limits of the shallow and the deep: green politics, philosophy, and praxis', *Environmental Politics*, 3, 369–94.

Barry, J. (1996), 'Sustainability, political judgement and citizenship', in B. Doherty and M. de Geus (eds) (1996b), *Democracy and Green Political Thought*. London and New York: Routledge.

Barry, J. (1999), *Rethinking Green Politics*. London: Sage.

Barry, J. and Wissenberg, M. (eds) (2001), *Sustaining Liberal Democracy*. Basingstoke and New York: Palgrave.

Beck, U. (1992), *Risk Society: Towards a New Modernity*. London: Sage.

Beck, U. (1995), *Ecological Politics in an Age of Risk*. Cambridge: Polity Press.

Beitz, C. R. (1979), *Political Theory and International Relations* (reissue with a new Afterword, 1999). Princeton, NJ: Princeton University Press.

Bellamy, R. (1999), *Liberalism and Pluralism*. London: Routledge.

Bellamy, R. and Jones, R. J. B. (2000), 'Globalization and democracy – an afterword', in B. Holden (ed.) (2000a), *Global Democracy: Key Debates*. London and New York: Routledge.

Benhabib, S. (1996), 'Toward a deliberative model of democratic legitimacy', in S. Benhabib (ed.), *Democracy and Difference*. Princeton, NJ: Princeton University Press.

Bienen, D., Rittberger, V. and Wagner, W. (1998), 'Democracy in the United Nations system', in D. Archibugi, D. Held and M. Köhler (1998b), *Re-imagining Political Community*. Cambridge: Polity Press.

Borg, S. (1998), 'Guarding intergenerational rights over natural resources', in E. Agius and S. Busuttil (eds) (1998), *Future Generations and International Law*. London: Earthscan.

Boutros-Ghali, B. (1996), *An Agenda for Democratization*. New York: United Nations. (A key section is reprinted as Chapter 7 in Holden (2000a).)

Brenton, T. (1994), *The Greening of Machiavelli*. London: Earthscan.

Bretherton, C. (1996), 'Introduction: global politics in the 1990s', in C. Bretherton and G. Ponton (eds), *Global Politics*. Oxford: Blackwell.

Brown, C. (1998), 'International social justice', in D. Boucher and P. Kelly (eds), *Social Justice: From Hume to Walzer*. London and New York: Routledge.

Brown, S. (1988), *New Forces, Old Forces, and the Future of World Politics*. Boston, MA: Little, Brown.

Brown, S. (1995), *New forces, Old Forces, and the Future of World Politics: Post-Cold War Edition*. New York: HarperCollins.

Budge, I. (1996), *The New Challenge of Direct Democracy*. Cambridge: Polity Press.

Burnheim, J. (1985), *Is Democracy Possible?* Cambridge: Polity Press.

Burnheim, J. (1986), 'Democracy, nation states and the world system', in D. Held and C. Pollitt (eds), *New Forms of Democracy*. London: Sage.

Caldwell, L. (1988), 'Beyond environmental diplomacy: the changing institutional structure of international cooperation', in J. E. Carroll (ed.), *International Environmental Diplomacy*. Cambridge: Cambridge University Press.

Camilleri, J. A. (1990), 'Rethinking sovereignty in a shrinking world', in R. J. B. Walker and S. H. Mendlovitz (eds), *Contending Sovereignties*. Boulder, CO and London: Lynne Rienner.

Camilleri, J. A. and Falk, J. (1992), *The End of Sovereignty?* Aldershot: Edward Elgar.

Cohen, J. L. and Arato, A. (1992), *Civil Society and Political Theory*. Cambridge, MA: MIT Press.

Commission on Global Governance, The (1995), *Our Global Neighbourhood*. Oxford: Oxford University Press.

Cooke, M. (2000), 'Five arguments for deliberative democracy', *Political Studies*, 48, 947–69.

Cox, R. (1997), Democracy in hard times: economic globalization and the limits to liberal democracy', in A. McGrew (ed.), *The Transformation of Democracy?* Cambridge and Milton Keynes: Polity Press in association with The Open University Press.

Crawford, J. (1994), *Democracy in International Law*. Cambridge: Cambridge University Press.

Czempiel, E.-O. (ed.), *Governance without Government: Order and Change in World Politics*. Cambridge: Cambridge University Press.

Dahl, R. A. (1989) *Democracy and its Critics*. New Haven, CT: Yale University Press.

Dahl, R. A. (1999), 'Can international organizations be democratic? A skeptic's view', in I. Shapiro and C. Hacker-Cordón (1999b), *Democracy's Edges*. Cambridge: Cambridge University Press.

Davidson, A. (2000), 'Democracy, class and citizenship in a globalizing world', in A. Vandenberg (ed.), *Citizenship and Democracy in a Global Era*. Basingstoke: Macmillan.

Delanty, G. (2000), *Citizenship in a Global Age*. Buckingham and Philadelphia, PA: Open University Press.

della Porter, D. and Kreisi, H. (eds) (1999), 'Social movements in a globalizing world: an introduction', in D. della Porter, H. Kriesi and D. Rucht, *Social Movements in a Globalizing World*. London and New York: Macmillan and St Martin's Press.

de-Shalit, A. (1995), *Why Posterity Matters*. London and New York: Routledge.

Dobson, A. (1995), *Green Political Thought*, 2nd edn. London and New York: Routledge.

Dobson, A. (1996), 'Democratising green political theory: preconditions and principles', in B. Dohery and M. de Geus (eds) (1996b), *Democracy and Green Political Thought*. London and New York: Routledge.

Dobson, A. (1998), *Justice and the Environment*. Oxford: Oxford University Press.

Dobson, A. (2000), *Green Political Thought*, 3rd edn. London and New York: Routledge.

Dobson, A. and Lucardie, P. (eds) (1993), *The Politics of Nature*. London and New York: Routledge.

Doherty, B. and de Geus, M. (1996a), 'Introduction', in B. Doherty and M. de Geus (eds) (1996b), *Democracy and Green Political Thought*. London and New York: Routledge.

Doherty, B. and de Geus, M. (eds) (1996b), *Democracy and Green Political Thought*. London and New York: Routledge.

Dryzek, J. S. (1987), *Rational Ecology: Environment and Political Economy*. Oxford: Blackwell.

Dryzek, J. S. (1990), *Discursive Democracy*. Cambridge: Cambridge University Press.

Dryzek, J. S. (1995), 'Ecology and discursive democracy: beyond liberal capitalism and the administrative state', in M. O'Connor (ed.), *Is Capitalism Sustainable? Political Economy and the Politics of Ecology*. New York and London: Guildford.

Dryzek, J. S. (1996), 'Strategies of ecological democratization', in W. M. Lafferty and J. Meadowcroft (eds) (1996b), *Democracy and the Environment*. Cheltenham: Edward Elgar.

Dryzek, J. S. (1999), 'Transnational democracy', *The Journal of Political Philosophy*, 7, 30–51.

Dunn, J. (1994a), 'Introduction: crisis of the nation state?', in J. Dunn (ed.) (1994b), *The Contemporary Crisis of the Nation State? Political Studies*, 42 (special issue).

Dunn, J. (ed.) (1994b), *The Contemporary Crisis of the Nation State? Political Studies*, 42 (special issue).

Earth Summit '92: The United Nations Conference on Environment and Development Rio de Janeiro 1992 (1992). London: The Regency Press.

Eckersley, R. (1992), *Environmentalism and Political Theory: Toward an Ecocentric Approach*. London: UCL Press.

Eckersley, R. (1996), 'Greening liberal democracy', in B. Doherty and M. de Geus (eds) (1996b), *Democracy and Green Political Thought*. London and New York: Routledge.

Economist, The (1999), 'Will NGOs democratise, or merely disrupt, global governance?', 11 December.

Elster, J. (1983), *Sour Grapes. Essays in the Subversion of Rationality*. Cambridge: Cambridge University Press.

Elster, J. and Slagstad, R. (eds) (1993), *Constitutionalism and Democracy*. Cambridge: Cambridge University Press.

Evans, T. (1997), 'Democratization and human rights', in A. McGrew (ed.) (1997b), *The Transformation of Democracy?* Cambridge and Milton Keynes: Polity Press in association with The Open University Press.

Falk, R. (1971), *This Endangered Planet: Prospects and Proposals for Human Survival*. New York: Vintage Books.

Falk, R. (1995), 'The world order between inter-state law and the law of humanity: the role of civil society institutions', in D. Archibugi and D. Held (eds) (1995), *Cosmopolitan Democracy*. Cambridge: Polity Press.

Falk, R. (1999), *Predatory Globalization*. Cambridge: Polity Press.

Falk, R. (2000), 'Global civil society and the democratic prospect', in B. Holden (ed.) (2000a), *Global Democracy: Key Debates*. London and New York: Routledge.

Feyerabend, P. K. (1975), *Against Method*. London: New Left Books.

Feyerabend, P. K. (1978), *Science in a Free Society*. London: New Left Books.

Feyerabend, P. K. (1987), *Farewell to Reason*. London: New Left Books.

Finger, M. (1994), 'NGOs and transformation: beyond social movement theory', in T. Princen and M. Finger (1994), *Environmental NGOs in World Politics*. London and New York: Routledge.

Fishkin, J. (1991), *Democracy and Deliberation*. New Haven, CT: Yale University Press.

Franck, T. M. (1992), 'The emerging right to democratic governance', *American Journal of International Law*, 86, 46–91.

Fukuyama, F. (1989), 'The end of history?', *The National Interest*, 16, 2–18.

Fukuyama, F. (1992), *The End of History and the Last Man*. London: Hamish Hamilton.

Galtung, J. (2000), 'Alternative models for global democracy', in B. Holden (ed.) (2000a), *Global Democracy: Key Debates*. London and New York: Routledge.

Gershman, C. (1993), 'The United Nations and the new world order', *Journal of Democracy*, 4, 5–16.

Giddens, A. (1991), *Modernity and Self-identity*. Cambridge: Polity Press.

Golub, J. (2000), 'Globalization, sovereignty and policy-making: insights from European integration', in B. Holden (ed.) (2000a), *Global Democracy: Key Debates*. London and New York: Routledge.

Goodin, R. E. (1996), 'Enfranchising the earth, and its alternatives', *Political Studies*, 44, 835–49.

Gore, A. (1992), *Earth in Balance*. New York/Boston/London: Houghton Mifflin.

Grubb, M., Koch, M., Munson, A., Sullivan, F. and Thomson, K. (1993), *The Earth Summit Agreements*. London: Earthscan.

Gutmann, A. and Thompson, D. (1996), *Democracy and Disagreement*. Cambridge, MA and London: Belknap Press.

Haas, P. M. (1992a), 'Obtaining international environmental protection through epistemic consensus', in I. H. Rowlands and M. Green (eds) (1992), *Global Environmental Change and International Relations*. London: Macmillan.

Haas, P. M. (ed.) (1992b), *Knowledge, Power, and International Policy Coordination*, special issue of *International Organization*, 46 (Winter).

Hadenius, A. (1997), 'Victory and crisis: introduction', in A.

Hadenius (ed.), *Democracy's Victory and Crisis*. Cambridge: Cambridge University Press.

Hardin, G. (1968), 'The tragedy of the commons', *Science*, 162, 1243–8.

Hardin, G. (1977), *The Limits to Altruism*. Indianapolis, IN: Indiana University Press.

Hardin, R. (1982), *Collective Action*. Baltimore, MD: Johns Hopkins University Press.

Harrison, R. (1993), *Democracy*. London: Routledge.

Hawken, P., Amory, B. L. and Lovens, L. H. (1999), *Natural Capitalism: the Next Industrial Revolution*. London: Earthscan.

Heater, D. (1990), *Citizenship: The Civic Ideal in World History, Politics and Education*. London and New York: Longman.

Heilbroner, R. L. (1974), *An Enquiry into the Human Prospect*. New York: Norton (2nd edn, 1980).

Held, D. (1995), *Democracy and the Global Order*. Cambridge: Polity Press.

Held, D. (2000), 'The changing contours of political community: rethinking democracy in the context of globalization', in B. Holden (2000a), *Global Democracy: Key Debates*. London and New York: Routledge.

Held, D., McGrew, A., Goldblatt, D. and Perraton, J. (1999), *Global Transformations*. Cambridge: Polity Press.

Higgot, R. and Payne, A. (2000), *The New Political Economy of Globalization*. Cheltenham: Edward Elgar.

Hindess, B. (1991), 'Imaginary presuppositions of democracy', *Economy and Society*, 20, 173–95.

Hirst, P. (1994), *Associative Democracy*. Cambridge: Polity Press.

Hirst, P. (1996), 'Democracy and civil society', in P. Hirst and S. Khilnani (eds), *Reinventing Democracy*. Oxford and Cambridge, MA: Blackwell.

Hirst, P. and Thompson, G. (1999), *Globalization in Question*, 2nd edn. Cambridge: Polity Press.

Hirst, P. and Thompson, G. (2000), 'Global myths and national policies', in B. Holden (ed.) (2000a), *Global Democracy: Key Debates*. London and New York: Routledge.

HMSO (2000), *Energy – The Changing Climate*. London: Her Majesty's Stationery Office.

Holden, B. (1974), *The Nature of Democracy*. London: Nelson.

Holden, B. (1988), 'New directions in democratic theory', *Political Studies*, 36, 324–33.

Holden, B. (1993), *Understanding Liberal Democracy*, 2nd edn. Hemel Hempstead: Harvester Wheatsheaf.

Holden, B. (1996a), 'Democratic theory and the problem of global warming', in B. Holden (ed.) (1996b), *The Ethical Dimensions of Global Change*. Basingstoke: Macmillan.

Holden, B. (1996b), *The Ethical Dimensions of Global Change*. Basingstoke: Macmillan.

Holden, B. (ed.) (2000a), *Global Democracy: Key Debates*. London and New York: Routledge.

Holden, B. (2000b), 'International social justice, global warming and global democracy', in A. Coates (ed.), *International Justice*. Aldershot: Ashgate Publishing.

Holden, B. (2000c), *Global Warming and 'Intergenerational Democracy'*. Reading University: Reading Papers in Politics, 22.

Holmes, S. (1993), 'Precommitment and the paradox of democracy', in J. Elster and R. Slagstad (eds) (1993), *Constitutionalism and Democracy*. Cambridge: Cambridge University Press.

Hont, I. (1994), 'The permanent crisis of a divided mankind: "Contemporary crisis of the nation state" in historical perspective', in J. Dunn (ed.) (1994b), *The Contemporary Crisis of the Nation State? Political Studies*, 42 (special issue).

Houghton, J. (1997), *Global Warming*. Cambridge: Cambridge University Press.

Hudock, A. C. (1999), *NGOs and Civil Society: Democracy by Proxy?* Cambridge: Polity Press.

Huntington, S. (1993), 'The clash of civilizations', *Foreign Affairs*, 72, 22–49.

Huntington, S. (1996), *The Clash of Civilizations and the Remaking of the World Order*. New York: Simon & Schuster.

Hurrell, A. and Kingsbury, B. (1992a), 'Introduction', in A. Hurrell and B. Kingsbury (eds) (1992b), *The International Politics of the Environment*. Oxford: Clarendon Press.

Hurrell, A. and Kingsbury, B. (eds) (1992b), *The International Politics of the Environment*. Oxford: Clarendon Press.

Hutchings, K. (1996), 'The idea of international citizenship', in B. Holden (1996b), *The Ethical Dimensions of Global Change*. Basingstoke: Macmillan.

Imber, M. (1997), 'Geo-governance without democracy? Reforming the UN system', in A. McGrew (ed.) (1997b), *The Transformation of Democracy?* Cambridge and Milton Keynes: Polity Press in association with The Open University Press.

IPCC (2001), *Climate Change 2001: The Scientific Basis*. Cambridge University Press.

Irwin, A. (1995), *Citizen Science*. London: Routledge.

Jacobs, M. (1997) 'Environmental valuation: deliberative democracy and public decision-making institutions', in J. Foster (ed.) *Valuing Nature*. London: Routledge.

Kavka, G. S. and Warren, V. (1983), 'Political representation for future generations', in R. Elliot and A. Gare (eds), *Environmental Philosophy*. Milton Keynes: The Open University Press.

Kelso, A. W. (1978), *American Democratic Theory: Pluralism and its Critics*. Westport, CT: Greenwood Press.

Kimber, R. (1989), 'On democracy', *Scandinavian Political Studies*, 12, 199–219.

Krasner, S. D. (ed.) (1983), *International Regimes*. Ithaca, NY: Cornell University Press.

Kuhn, T. S. (1962, 1970), *The Structure of Scientific Revolutions* (1st edn 1962, 2nd edn 1970). Chicago, IL: University of Chicago Press.

Lafferty, W. M. and Meadowcroft, J. (1996a), 'Democracy and the environment: congruence and conflict', in W. M. Lafferty and J. Meadowcroft (eds) (1996b), *Democracy and the Environment*. Cheltenham: Edward Elgar.

Lafferty, W. M. and Meadowcroft, J. (eds) (1996b), *Democracy and the Environment*. Cheltenham: Edward Elgar.

Leet, M. (1998), 'Jürgen Habermas and deliberative democracy', in A. Carter and G. Stokes (eds), *Liberal Democracy and Its Critics*. Cambridge: Cambridge University Press.

Lindsay, A. D. (1943), *The Modern Democratic State*. Oxford: Oxford University Press.

Linklater, A. (1998), 'Citizenship and sovereignty in the post-Westphalian European state', in D. Archibugi, D. Held and M. Köhler (eds) (1998b), *Re-imagining Political Community*. Cambridge: Polity Press.

Lipschutz, R. D. (1996), *Global Civil Society and Global Environmental Governance*. Albany, NY: State University of New York Press.

List, M. and Rittberger, V. (1992), 'Regime theory and international environmental management', in A. Hurrell and B. Kingsbury (eds) (1992b), *The International Politics of the Environment*. Oxford: Clarendon Press.

Lomborg, B. (2001), *The Skeptical Environmentalist: Measuring the Real State of the World*. Cambridge: Cambridge University Press.

McGrew, A. (1997a), 'Democracy beyond borders?: globalization and the reonstruction of democratic theory and politics', in A. McGrew (1997b), *The Transformation of Democracy?* Cambridge and Milton Keynes: Polity Press in association with The Open University Press.

McGrew, A. (ed.) (1997b), *The Transformation of Democracy?* Cambridge and Milton Keynes: Polity Press in association with The Open University Press.

Macpherson, C. B. (1977), *The Life and Times of Liberal Democracy*. Oxford: Oxford University Press.

Magnusson, W. (1990), 'The reification of political community', in R. J. B. Walker and S. H. Mendlovitz (eds) (1990b), *Contending Sovereignties*. Boulder, CO and London: Lynne Rienner.

Malhotra, A. (1998), 'A commentary on the status of future generations as a subject of international law', in E. Agius and S. Busuttil (eds) (1998), *Future Generations and International Law*. London: Earthscan.

May, J. D. (1978), 'Defining democracy: a bid for coherence and consensus', *Political Studies*, 26, 1–14.

Micklethwait, J. and Wooldridge, A. (2000), *A Future Perfect: The Challenge and Hidden Promise of Globalisation*. London: Heinemann.

Mill, J. S. [1861] (1958), *Considerations on Representative Government* (ed. C. V. Shields). Indianapolis, IN: Bobbs-Merrill.

Miller, D. (1992), 'Deliberative democracy and social choice', *Political Studies*, 40 (special issue, *Prospects for Democracy*), 54–67.

Mills, M. (1996), 'Green democracy: the search for an ethical solution', in B. Doherty and M. de Geus (eds) (1996b), *Democracy and Green Political Thought*. London and New York: Routledge.

Mueller, D. (1979), *Public Choice*. Cambridge: Cambridge University Press.

Mulgan, G. J. (1994), *Politics in an Antipolitical Age*. Cambridge: Polity Press.

Naess, A. (1989), *Ecology, Community and Lifestyle*. Cambridge: Cambridge University Press.

Nardin, T. (1983), *Law, Morality and the Relations of States*. Princeton, NJ: Princeton University Press.

O'Connor, J. (1990), 'Commentary', Workshop on Ecology, Committee on the Political Economy of the Good Society, American Political Science Association, San Francisco, CA.

Ophuls, W. (1977), *Ecology and the Politics of Scarcity*. San Francisco, CA: Freeman.

Oppenheimer, M. (1990), 'Responding to climate change: the

crucial role of the NGOs', in H.-J. Karpe, D. Otten and S. C. Trinidade (eds), *Climate and Development*. New York/Berlin/Heidelberg: Springer-Verlag.

O'Riordan, T. (1981), *Environmentalism*. London: Pion.

Ostrom, E. (1990), *Governing the Commons: The Evolution of Institutions for Collective Action*. Cambridge: Cambridge University Press.

Oye, K. A. (ed.) (1986), *Cooperation under Anarchy*. Princeton, NJ: Princeton University Press.

Paehlke, R. (1988), 'Democracy, bureaucracy and environmentalism', *Environmental Ethics*, 10, 291–308.

Paehlke, R. (1996), 'Environmental challenges to democratic practice', in N. M. Lafferty and J. Meadowcroft (eds) (1996b), *Democracy and the Environment*. Cheltenham: Edward Elgar.

Parekh, B. (ed.) (1973), *Bentham's Political Thought*. London: Croom Helm.

Pateman, C. (1985), *The Problem of Political Obligation: A Critique of Liberal Theory*. Cambridge: Polity Press.

Paterson, M. (1992), 'Global warming', in C. Thomas, *The Environment in International Relations*. London: RIIA.

Paterson, M. (1996), 'International justice and global warming', in B. Holden (ed.) (1996), *The Ethical Dimensions of Global Change*. Basingstoke: Macmillan.

Pearce, D. (1992), 'Economics and global environmental change', in I. H. Rowlands and M. Green (eds) (1992), *Global Environmental Change and International Relations*. London: Macmillan.

Pearce, D. and Barbier, E. (2000), *Blueprint for a Sustainable Economy*. London: Earthscan.

Perraton, J. (2000), 'Hirst and Thompson's "Global myths and national policies": a reply', in B. Holden (ed.) (2000a), *Global Democracy: Key Debates*. London and New York: Routledge.

Pitkin, H. L. (1967), *The Concept of Representation*. Berkeley, CA/Cambridge: University of California Press/Cambridge University Press.

Portney, P. R. and Weynant, J. P. (1999a), 'Introduction', in P. R. Portney and J. P. Weynant (eds) (1999b), *Discounting and Intergenerational Equity*. Washington, DC: Resources for the Future.

Portney, P. R. and Weynant, J. P. (eds) (1999b), *Discounting and Intergenerational Equity*. Washington, DC: Resources for the Future.

Princen, T. and Finger, M. (1994), *Environmental NGOs in World Politics*. London and New York: Routledge.

Ricci, D. M. (1971), *Community Power and Democratic Theory*. New York: Random House.

Rittberger, V. and Mayer, P. (eds) (1993), *Regime Theory and International Relations*. Oxford: Clarendon Press.

Rosenau, J. N. (1992), 'Governance, order and change in world politics', in J. N. Rosenau and E.-O. Czempiel (eds), *Governance without Government: Order and Change in World Politics*. Cambridge: Cambridge University Press.

Rowlands, I. H. (1995), *The Politics of Global Atmospheric Change*. Manchester: Manchester University Press.

Rowlands, I. H. and Green, M. (eds) (1992), *Global Environmental Change and International Relations*. London: Macmillan.

Sagoff, M. (1988), *The Economy of the Earth: Philosophy, Law and the Environment*. Cambridge: Cambridge University Press.

Sands, P. (1998), 'Protecting future generations: precedents and practicalities', in E. Agius and S. Busuttil (eds) (1998), *Future Generations and International Law*. London: Earthscan.

Sartori, G. (1987), *The Theory of Democracy Revisited*. Chatham, NJ: Chatham House.

Saward, M. (1993), 'Green democracy?', in A. Dobson and P. Lucardie (eds) (1993), *The Politics of Nature*. London and New York: Routledge.

Saward, M. (1996), 'Must democrats be environmentalists?', in B. Doherty and M. de Geus (eds) (1996b), *Democracy and Green Political Thought*. London and New York: Routledge.

Saward, M. (1998), *The Terms of Democracy*. Cambridge: Polity Press.

Saward, M. (2000), 'A critique of Held', in B. Holden (ed.) (2000a), *Global Democracy: Key Debates*. London and New York: Routledge.

Schelling, T. C. (1999), 'Intergenerational discounting', in P. R. Portney and J. P. Weynant (eds) (1999b), *Discounting and Intergenerational Equity*. Washington, DC: Resources for the Future.

Schiffler, W. (1954), *The Legal Community of Mankind*. New York: Columbia University Press.

Schumacher, E. (1976), *Small Is Beautiful*. London: Sphere.

Segall, J. D. (1997), 'A first step for peaceful cosmopolitan democracy', *Peace Review*, 9, 337–44.

Sen, A. and Williams, B. (eds) (1982), *Utilitarianism and Beyond*. Cambridge: Cambridge University Press.

Shapiro, I. and Hacker-Cordón, C. (1999a), 'Outer edges and inner edges', in I. Shapiro and C. Hacker-Cordón (1999b), *Democracy's Edges*. Cambridge: Cambridge University Press.

Shapiro, I. and Hacker-Cordón, C. (eds), (1999b), *Democracy's Edges*. Cambridge: Cambridge University Press.

Shaw, M. (1994), *Global Society and International Relations*. Cambridge: Polity Press.

Shue, H. (1992), 'The unavoidability of justice', in A. Hurrell and B. Kingsbury (eds) (1992b), *The International Politics of the Environment*. Oxford: Clarendon Press.

Sikora, R. I. and Barry, B. (1978a), 'Introduction', in R. I. Sikora and B. Barry (eds) (1978b), *Obligations to Future Generations*. Philadelphia, PA: Temple University Press.

Sikora, R. I. and Barry, B. (eds) (1978b), *Obligations to Future Generations*. Philadelphia, PA: Temple University Press.

Smith, G. and Wales, C. (2000), 'Citizens' juries and deliberative democracy', *Political Studies*, 48, 1: 51–65.

Smith, H. (2000), 'Why is there no international democratic theory', in H. Smith (ed.), *Democracy and International Relations*. Basingstoke: Macmillan.

Spitz, E. (1979), 'Defining democracy: a nonecumenical reply to May', *Political Studies*, 27, 127–8.

Stein, T. (1998), 'Does the constitutional and democratic system work? The ecological crisis as a challenge to the political order of constitutional democracy', *Constellations*, 4, 420–49.

Stone, C. D. (1998), 'Safeguarding future generations', in E. Agius and S. Busuttil (eds) (1998), *Future Generations and International Law*. London: Earthscan.

Talmon, J. L. (1952), *The Origins of Totalitarian Democracy*. London: Secker & Warburg.

Tarrow, S. (1994), *Power in Movement: Social Movements, Collective Action and Politics*. Cambridge: Cambridge University Press.

Taylor, B. P. (1996), 'Democracy and environmental ethics', in W. M. Lafferty and J. Meadowcroft (eds) (1996b), *Democracy and the Environment*. Cheltenham: Edward Elgar.

Taylor, M. (1987), *The Possibility of Cooperation*. Cambridge: Cambridge University Press.

Thomas, C. (1992), *The Environment and International Relations*. London: RIIA.

Thompson, J. (1998), 'Community identity and world citizenship', in D. Archibugi, D. Held and M. Köhler (1998), *Re-imagining Political Community*. Cambridge, Polity Press.

Walker, R. J. B. and Mendlovitz, S. H. (1990a), 'Interrogating state sovereignty', in R. J. B. Walker and S. M. Mendlovitz (eds)

(1990b), *Contending Sovereignties*. Boulder, CO and London: Lynne Rienner.

Walker, R. J. B. and Mendlovitz, S. H. (eds) (1990b), *Contending Sovereignties*. Boulder, CO and London: Lynne Rienner.

Walzer, M. (1992), 'The civil society argument', in C. Mouffe (ed.), *Dimensions of Radical Democracy*. London: Verso.

Walzer, M. (ed.) (1995), *Toward a Global Civil Society*. Providence, RI: Berghahn Books.

Wapner, P. (1996), *Environmental Activism and World Civic Politics*. Albany, NY: State University of New York Press.

Ward, H. (1998), 'State, association, and community in a sustainable, democratic polity: towards a green associationalism', in F. H. J. M. Coenen, D. Huitema and L. J. O'Toole, Jr (eds), *Participation and the Quality of Environmental Decision Making*. Dordecht/Boston/London: Kluwer Academic Publishers.

Ware, A. (1992), 'Liberal democracy: one form or many', in D. Held (ed.), *Prospects for Democracy, Political Studies*, 40 (special issue).

WCED (World Commission on Environment and Development) (1987), *Our Common Future*. Oxford: Oxford University Press.

Weale, A. (1992), *The New Politics of Pollution*. Manchester and New York: Manchester University Press.

Weale, A. (1999), *Democracy*. London: Macmillan.

Westra, L. (1994), *An Environmental Proposal for Ethics*. Lanham, MD: Rowman & Littlefield.

Whelan, F. G. (1983), 'Prologue: democratic theory and the boundary problem', in J. R. Pennock and J. W. Chapman (eds), *Liberal Democracy* (Nomos XXV). New York: New York University Press.

Wolff, R. P. (1970), *In Defence of Anarchism*. New York: Harper & Row.

Wollheim, R. (1958), 'Democracy', *Journal of the History of Ideas*, 19, 225–42.

Wollheim, R. (1962), 'A paradox in the theory of democracy', in P. Laslett and G. Runciman (eds), *Essays in Politics, Philosophy and Society*, second series. Oxford: Blackwell.

Young, O. R. (1994), *International Governance: Protecting the Environment in a Stateless Society*. Ithaca NY and London: Cornell University Press.

Zolo, D. (2000), 'The lords of peace: from the Holy Alliance to the new international criminal tribunals', in B. Holden (ed.) (2000a), *Global Democracy: Key Debates*. London and New York: Routledge.

Index